Home Plumbing
Manual

Andy Blackwell

Contents

1 Introduction   **6**

About this book   7
A brief history of plumbing   8
What can we do for ourselves?   10
What can't we do?   10
Water regulations   12

2 Know your home   **14**

The most important taps in your home   15
Where is my mains stop tap?   15
Where is the outdoor stopcock   15

Identifying hot- and cold-water systems   17
Stored systems   17
Mains-fed systems   17
What cold-water system do you have in your home?   17
What hot-water system do you have in your home?   18
Finding your cold-water storage tank   21
Going on holiday   22

What type of central heating do you have?   23
Gravity-fed central heating   23
High-pressure central heating   25
Draining down your central heating   27
Refilling your central heating   29
Words of advice   29

3 Plumbing wear, tools and materials   **30**

Essential safety equipment   31
Planning   32
Essential tools   32

4 Dealing with emergencies   **36**

What to do in an emergency   37
Causes of leaks   37
Burst pipes   38
Frozen pipes   42

5 Insulating pipework and storage tanks   **44**

Insulating your pipework   45
Applying foam insulation to pipework   46
Insulating your storage tanks   48
Fitting a Regulation 16 kit (Byelaw 30 kit)   49

6 Pipework and basic techniques   **52**

Plastic push-fit pipework   53
Tectite metal push-fit pipework   55
Copper compression pipework   56
Using jointing compounds and tape   58
Removing compression fittings   58
Using compression joints with plastic pipework   59
Using chrome pipe   60
Soldered pipework   60
Black iron or steel pipework   63
Lead pipework   64
Waste pipework   65
Name that fitting   68
Bending pipework   70
Pipe sizing   72
Supporting pipework   73

7 Maintaining and repairing emergency valves   **74**

The water company's stop tap   75
The mains stop tap (or stopcock)   75
The hot-water stop tap   78
Isolation valves   79

Home
Plumbing
Manual

First published in May 2012

British Library Cataloguing in Publication Data
A catalogue record for this book is available from the British Library.

ISBN 978 0 85733 069 7

Published by Haynes Publishing,
Sparkford, Yeovil, Somerset BA22 7JJ, UK
Tel: 01963 442030 Fax: 01963 440001
Int. tel: +44 1963 442030 Int. fax: +44 1963 440001
E-mail: sales@haynes.co.uk
Website: www.haynes.co.uk

Haynes North America Inc.
861 Lawrence Drive, Newbury Park,
California 91320, USA

Printed in the USA by Odcombe Press LP,
1299 Bridgestone Parkway, La Vergne, TN 37086

Credits

Author:	Andy Blackwell
Project Manager:	Louise McIntyre
Copy editor:	Ian Heath
Page design:	James Robertson
Illustrator:	Dominic Stickland
Index:	Matthew Gale

Author acknowledgements

There were a host of people who supplied advice and help during the time it took me to write this manual. If I haven't mentioned you I'm sorry for the omission but it doesn't mean I didn't appreciate the help you gave.

■ Thanks to my business partner Neil 'Sugs' Hayes for all his help with the photos and a big thanks to all the customers of A1 Perfect Plumbing who didn't object to me wandering around their homes taking photos when I should have been working.

■ A big thanks to Ian Holt from Ian Holt Services (IHS) of Grantham for all his help checking for technical omissions and errors. I doubt we spotted them all but there are now far fewer than there might have been.

■ To my mate Tim Russell for all his help with proof reading.

■ To Claire Jennings and Beth Barrett at Bristan.

■ To Clare Campbell and Dale Banks at Dimplex renewables.

■ To Astrid and Lutz at Aquality.

■ To all the plumbing merchants of Grantham who let me 'borrow' plumbing items on an almost continuous basis, especially Mat, Mick, Kenny, Chris and Allan at PTS; Chris, Mat, Garry and Jake at The Plumb Centre; Simon, Dave, Lee and Bryon at Grahams and John at Travis Perkins.

■ To Dave and the 'Men of Metal' at Bullen's (aka Lincolnshire processed scrap metal Recycling Co) for all their help in sourcing the more obscure and ancient plumbing items.

■ And of course a huge thanks to Louise McIntyre of Haynes Publishing for giving me this opportunity.

■ Finally a thanks to my wife Leanne for all her support and for tolerating all the evenings and weekends that I spent typing away whilst she managed our home and tackled our baby boy. And to my baby boy, Marty, for not keeping his dad up all night, every night.

11	**In the bathroom**	**132**
	Using sealant	133
	A (very) brief guide to tiling	134
	Replacing washbasin and bath taps	135
	Removing an old toilet	137
	Fitting a new toilet	139
	Removing a bath	146
	Fitting a new bath	147
	Removing a washbasin	157
	Fitting a new washbasin	158
	Fitting a new shower	164
	Fitting a bidet	169
	Fitting a shower tray and cubicle	170
	Swapping your bath for a shower tray	171
	Wet rooms	171

12	**Central heating and hot water**	**172**
	Bleeding a radiator	173
	Maintaining your central-heating system	173
	Keeping your central-heating system clean	177
	Removing a radiator	179
	Fitting a new radiator	181
	Fitting new radiator valves	184
	Fitting a heated towel rail	185
	Common central-heating problems	186
	Replacing the central-heating pump	189
	Replacing the motorised valve	191
	Fitting a new hot-water cylinder	192
	Replacing an immersion heater	197
	Boilers	199

13	**Going green**	**202**
	Insulation	203
	Energy neutral homes	203
	Underfloor heating	203
	Solar thermal energy	209
	Heat pumps	212
	Biomass boilers	215
	Rainwater harvesting	216

Glossary	**220**

Useful contacts	**223**

Index	**224**

8	**Basic plumbing: the usual suspects**	**82**
	Help, I have a leak!	83
	Leaking compression joints	83
	Leaking washers	84
	My sink is blocked	85
	My toilet is blocked	88
	Blocked drains	89
	My toilet won't flush properly	91
	My toilet is overflowing	97
	The siphon is leaking	100
	The tanks in the loft are overflowing	101
	My taps are dripping	102
	Dealing with an airlock	107
	Fitting a new cold-water storage tank	108

9	**In the garden**	**112**
	Fitting an outside tap	113
	Garden irrigation systems	116
	Fitting a water butt	117

10	**In the kitchen**	**118**
	Fitting new monobloc taps	119
	Fitting a new kitchen sink	121
	Fitting a new dishwasher or washing machine	124
	Fitting a fridge ice-maker	129
	Fitting a water filter	129
	Fitting a water softener	129

Home Plumbing Manual

1 INTRODUCTION

About this book	7
A brief history of plumbing	8
What can we do for ourselves?	10
What can't we do?	10
Water regulations	12

About this book

It's possible to live your entire life in a house that never has a plumbing problem. It's also possible to win the lottery three times on the trot. I dare say it might happen one day, but I very much doubt that it'll happen to you or I. So in the meantime let's just assume that sooner or later we'll crash headlong into a plumbing problem. These can be as mild as a dripping tap or as monstrous as a flooded house; they can be dangerous, complicated and best left to the professionals, or they can be so simple and straightforward that you might as well solve them yourself – if only you had a bit more knowledge of the subject.

And that's where this book comes in. We're going to journey together through the wondrous world of plumbing, from the private lives of leaks and drips, right through to the splendour of bathroom installations and solar-powered hot water. Some of you might just want to know how to fit an outside tap, others might want to try their hand at plumbing in all its glory; others may not want to undertake the work at all, but just want to know what's involved so that they can choose and evaluate the right professional. Whatever your needs, if they're plumbing-related this is the book for you.

Many people have the notion that plumbing is beyond them, not because they lack knowledge but because they lack strength. Fortunately for me this isn't the case. Most aspects of plumbing require little or no strength, and for those few areas where a bit of brute force can come in handy there are tools galore to help those of us who don't spend our days down the gym. That said, most plumbing work comes under the heading of 'moderate exercise', and it's possible to injure yourself at the start by overdoing things a bit. The most common injuries for the beginner are tennis elbow and housemaids' knee. Fortunately, however, most DIY stores sell knee-pads and electric screwdrivers, and you'd do well to purchase these items before you even think about any plumbing.

WELL...AT LEAST WE KNOW THE BASEMENT DOESN'T LEAK.

A brief history of plumbing

Gather enough people together, cram them into a relatively small area, throw a few parties, a feast or two and the odd drinking competition, and sooner or later someone – even if it's a crowd of teenagers – will say 'This place is a tip! It smells, it's unhealthy, I keep treading in stuff I'd rather not tread in, we need to sort this out' – and thus was born the plumber.

By all accounts the Egyptians were the first to realise the benefits of sewage systems and clean water, but the news that it was possible to live in a city, enjoy yourself and not wake up smelling of poo soon caught on. Before you knew it, you couldn't call yourself civilised unless you had at least one plumber in town – there was probably even a civil engineer or two hanging around the place, but whilst they sat around in togas eating grapes and taking all the glory, who was actually doing the work? Yup, the plumber ... and possibly the brickie ... and maybe a few others.

Sadly, in Northern Europe it took a little longer for the idea of sanitation and sewerage to catch on. This might have been because togas showed up the dirt more, or possibly because freezing northern winters tended to kill off both the germs and the smell. Whatever the reason, it wasn't until the late 1800s that Londoners finally got fed up with cholera and typhoid

FLEET-STREET.—DEEPENING THE SEWER.

routinely decimating the population and decided to build a municipal freshwater and sewage system.

This worked so well at obliterating waterborne disease that we now largely take our plumbing systems for granted. The chances of Auntie Mabel dying from typhoid within the urban hell-hole that is Guildford are now so remote that we completely forget that most of the world still suffers from such diseases and can only dream of 'on-tap', fresh, clean water and invisible and efficient sewage systems that silently run the waste products of our lives far away where they can do no harm.

This is something to bear in mind the next time you're working yourself into a froth just because your tap drips; thank your lucky stars that at least the water coming out of that tap isn't going to kill you.

Sadly, although plumbers created civilisation and saved the lives of millions, few if any have entered the history books. The sole exception to this is Thomas Crapper, the inventor of the toilet ...

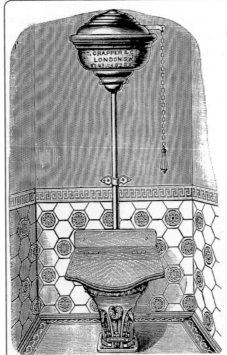

except, of course, he wasn't. Apparently Mr Crapper grew to fame because he was a purveyor of fine toilets to the nobility – although he did find time to invent the ballcock. Sadder still is the fact that the flushing loo wasn't invented by someone with an equally humorous name instead of boring old John Harrington – which may explain the term 'off to the John' but probably doesn't. According to Wikipedia, Mr Harrington invented the flushing toilet in 1596 but it wasn't deemed 'practicable' for another 182 years, when Joseph Bramah updated the design, presumably by putting a hole in the bottom of it.

Anyway, enough of history. Suffice to say that plumbers, regardless of name and fame, have been saving the day for countless generations, and now's your chance to join that merry band.

" This must be the place ! "

What can we do for ourselves?

If you make a mistake when working with water you're liable to get very wet. If you make a mistake with oil you'll end up smelling of kerosene for a week, you might set fire to your house, and you could possibly poison half of the Thames. If you make a mistake with gas you could end up scattered over four counties before you can say 'Oops!'

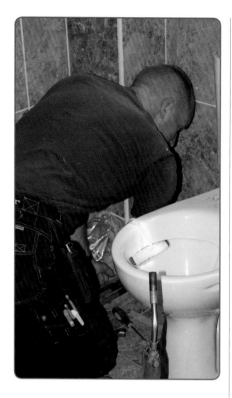

From these brief descriptions you might have gathered that we're going to stick to water, partly because you need professional qualifications and membership of recognised bodies to deal with gas, but mainly because we value our lives.

Anyone can work with water. In fact anyone can call themselves a plumber, and this is half the trouble we have in the UK, with so called 'rogue tradesmen' – who may well be rogues but certainly aren't qualified tradesmen. However, whilst water might only occasionally blow you up – see 'Unvented hot-water cylinders' below – it's still possible to cause floods of almost biblical proportions and to poison half the neighbourhood if you do things wrong.

In an attempt to stop such events occurring the building and water regulations were invented, and whether you're a homeowner, my-mate's-mate or a professional plumber you ignore these rules at your financial peril.

What can't we do?

GAS WORK

To do anything with a gas pipe, other than lovingly polish it, you need to be a member of the Gas Safe register, which in turn means that you have to be a fully qualified gas engineer. The Gas Safe register took over from Corgi in April 2009 (2010 for Northern Ireland), and is the body that

deals with gas safety. If you're not registered with the gas safe register it's illegal for you to work with gas .... unless you live in any of the channel isles that isn't Guernsey. Yup, for some inexplicable reason anyone can play with gas in the majority of the channel isles. In fairness they do seem to prefer you to be Gas Safe registered but they don't insist. So, if you ever come across a small crater whilst visiting Jersey you can at least hazard a guess at its cause.

OIL

At the moment (things will probably change in the near future) anyone can fit or service an oil boiler provided they inform the relevant building authority before they start, and the work is subsequently

An old oil tank conforming to few, if any, current regs.

inspected to ensure it meets the required standards. However, whilst it's difficult to kill yourself when working with oil, it's not impossible. It also takes skill and some very expensive equipment to service and test your boiler to make sure it's running correctly and efficiently. Finally, one wrong move could result in you poisoning your local water supply, which will normally result in a fine of eye-watering proportions.

With all this in mind you really ought to leave oil boilers to the professionals, in other words OFTEC-registered technicians.

UNVENTED HOT-WATER CYLINDERS

As is often the case, the unvented hot-water cylinder was invented by an Englishman, in this case Sir Thomas Hawksley. In the dark, distant year of 1861 Sir Thomas had hardly begun to extol the benefits of his new cylinder when it was banned in the UK, for the simple reason that it was, potentially, a bomb. Bizarrely enough this fact didn't seem to deter people of a foreign persuasion, and the unvented cylinder was a roaring success abroad, so much so that the UK government finally relented and let it back into the UK a mere 128 years after banning it. However, the fact that it could be a bomb hasn't changed.

In an unvented cylinder the water is stored at about 3bar – three times atmospheric pressure – which, in itself, is perfectly safe. When things start getting 'interesting' is when you start heating the water up. In normal circumstances water will start to boil at 100°C. However, if the water is being stored under pressure, *ie* 3bar, it doesn't – in fact it doesn't even begin to think about boiling until about 132°C.

Now, this doesn't sound much of a difference, and it isn't ... until the container that the water's in cracks. If this happens the water that was a liquid at 3bar pressure immediately becomes steam at 1bar pressure. Again, this doesn't sound like a big deal ... until you realise that steam requires a container 1,600 times bigger than the one it currently occupies – and it wants it NOW! The result of this insistence is a very big explosion.

To stop this happening, an unvented cylinder comes with an array of safety devices: if one fails, another kicks in, if this fails, yet another kicks in. The problem is that safety devices are only safe if they've been installed properly and serviced regularly, and for this reason the homeowner needs to stay well clear of them. If you do have an unvented cylinder (see below for indentifying features), arrange for a G3-qualified plumber to service it every year. In the interim, keep an eye on it yourself. If you ever see water dripping or running out of the pipes connected to the cylinder, call a G3-qualified plumber immediately to check it out.

NOTE

'G3' refers to part G of the building regulations, section 3, which deals with hot water storage and the dangers inherent in storing hot water at high pressure. To minimise these dangers the building regulations stipulate that only a 'competent' person should work on such systems, meaning someone who's completed a course on unvented cylinders; hence the term 'G3-qualified'.

Identifying an unvented cylinder

The easiest way to indentify your hot-water cylinder is by discovering what it isn't, which means trying to find the vent pipe and failing.

A vented hot-water cylinder is normally made from copper and covered in either fibreglass lagging or a foam insulating material – more often than not this is green.

At the top of the cylinder is a single pipe that emerges and then immediately turns sideways so that it's more or less horizontal to the tank but actually rising very gently.

This pipe then splits into a tee, with one pipe going up and the other going down.

If you follow the pipe going up you'll eventually arrive at a cold-water storage tank, and this 'vent' pipe will terminate over the top of it. It will be

arranged so that if water ever fills this pipe it will just pour out of it and into the cold-water storage tank.

If you managed to find this vent pipe then you have a standard 'vented' hot-water cylinder. If you didn't, the odds are you have an 'unvented cylinder', which have the following general characteristics:

- You usually can't see the insulation, *ie* it has an outer skin that's usually made from thin steel.

- It either has a flat top or there's an expansion vessel close by.

- It has a series of valves attached to it, or to the expansion vessel.

- It has a 'tundish' close by – a gap in the pipe so that you can see the water running through it. Why should you want to see the water? Well, if all is well there should never be water in this pipe. However, if any of the safety devices kick in you'll see water either dripping or pouring through it. When this happens something's gone wrong and you need to call in a plumber with the relevant 'G3' qualifications to check the cylinder.

If you're sure you have an unvented cylinder then you need to get it checked once a year by a suitably qualified plumber to ensure all is well.

If you've followed the above guidance and still aren't sure what you have, then it might be a good idea to call a plumber out and have them verify the type of cylinder.

Water regulations

Water regulations are immensely important and apply to all homeowners and anyone who wants to do any work with water. Sadly, like all regulations they're also astonishingly dull.

That said, we'll be referring to the water regs throughout this book, and for those of you who are going to undertake larger plumbing projects it's essential that you become familiar with them. You can buy the full guide – imaginatively called the *Water Regulations Guide* – from most online bookstores. If you search around online you can also find some free introductory guides that are worth looking at – the Northumbrian Water

one by J. McClean has some classic photos of 'what not to do' (http://www.nwl.co.uk/Water_regulations.pdf).

For those of you who have no intention of reading the water regulations in full I'll attempt to outline the basic tenets without going into them too deeply.

The primary areas covered are waste, undue consumption, misuse, erroneous measurement and contamination.

WASTE, UNDUE CONSUMPTION AND MISUSE

These three are in many respects different aspects of the same issue, namely failure to treat water as a valuable and rare resource.

This all seems to stem from the fact that it's been known to rain in the UK – sometimes quite a bit. In fact it rains so frequently that most of us have grown up with the notion that our verdant isles are overly blessed with water.

Sadly, this is a myth. I won't dispute that we get a fair amount of rain, but the fact is that it falls on an awful lot of people, all of whom like to wash themselves, their cars, their gardens, their dishes, their floors. We use it to flush away our waste, we use it to blast our driveways clean, and we've even been known to drink it.

The truth is that, per person, the UK has less rainfall than some areas of the Sudan, yet we still treat it as an essentially worthless, commonplace commodity for the simple reason that it seems to be falling out of the sky almost constantly.

To counter this misconception the water regulations are a legal obligation not to *waste* water, *ie* whatever water device you install, whether it be a tap, a bath, a shower etc, neither it nor its pipework will waste water by leaking. The easiest way to be sure of this is to only buy items that are WRAS (Water Regulations Advisory Scheme) approved, then make sure they're installed correctly and tested for leaks.

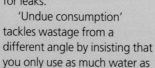

'Undue consumption' tackles wastage from a different angle by insisting that you only use as much water as needed. For example, in days gone by it used to take nine litres of water to flush a toilet and, as a result, going to the loo made up a large proportion of a household's water usage. To combat this the water regulations insist that new toilets are designed to have a maximum flush of just six litres and to have a half-flush capability.

'Misuse' could just as easily have been entitled 'Don't be stupid'. In other words, if you're going to build an Olympic-sized swimming pool at the bottom of your garden, then at least have the common decency to let the water board know where all their water's just disappeared to.

ERRONEOUS MEASUREMENT

'Erroneous measurement' is just a polite way of saying 'Don't nick our water!' *ie* don't try to circumvent the water meter or make the meter give a wrong reading etc.

CONTAMINATION

This is the big one. All the water coming into the home via the water mains is of drinking quality, and the water authorities go to an awful lot of trouble to ensure this – even though they know that most of the water won't be used for drinking or even cooking. As a result they're apt to get a bit miffed if some idiot contaminates their water.

The most common route for contamination is known as 'back-flow'. What does this mean? Well, in normal circumstances the cold water coming into your house is under a fair amount of pressure. Not only does this speed up the filling of your bath or the watering of your garden but it also ensures that water is only flowing in one direction, *ie* out of the hosepipe and on to your lawn. In this arrangement little or nothing can contaminate the mains cold-water pipework because little or nothing can get into it. But what happens if the mains pressure was to suddenly drop?

Let's suppose that a builder has just dug straight through the mains pipe two miles down the road from your home. Suddenly you find yourself with little or no water. So you drop the hose and wander back to the house looking for kinks in the hose. While you're doing this the water on your garden suddenly finds itself getting sucked back into the now empty hose and, eventually, back into the mains cold-water supply. Sadly this water is mixed with dog dirt, pesticides, plant food and God only knows what else. Yes, it's the builders' fault that the mains has lost its pressure, but it's still your fault that you've just contaminated the water supply and poisoned half your neighbourhood.

To prevent such scenarios – and they happen with alarming regularity – the water regulations insist that you have devices for preventing back-flow. In the case of the hosepipe example, you should have what is called a 'double-check' valve on your outside tap. This is a simple device that will only ever allow the water to flow one way. If this is fitted and you lose pressure, it doesn't matter because your tap won't let water flow back into the mains supply.

TIP *Since some taps have check valves and some don't, many hose manufacturers now supply hoses with a check valve built in. This doesn't mean you don't need one on the tap – it just means you're miles away from a fine if anything goes amiss with your mains water.*

So that's the water regulations. Yes, they're a bit dull but what I'm trying to impress upon you is that they aren't just mindless bureaucracy. They address a genuine need, and if you ignore them, whether you're the person doing the work or the homeowner employing that person, you're both going to get into trouble, not only with the water company but probably also with the people living nearby who've spent the entire weekend staring into their lavatory and shouting 'Barf...'.

2 KNOW YOUR HOME

The most important taps in your home   15

Where is my mains stop-tap?   15
Where is the outdoor stopcock?   15

Identifying hot- and cold-water systems   17

Stored systems   17
Mains-fed systems   17
What cold-water system do you have in your home?   17
What hot-water system do you have in your home?   18
Finding your cold-water storage tank   21
Going on holiday   22

What type of central heating do you have?   23

Gravity-fed central heating   23
High-pressure central heating   25
Draining down your central heating   27
Refilling your central heating   29
Words of advice   29

The most important taps in your home

Where is my mains stop tap?

The most important tap in your home is the 'mains stop tap'. As a general rule it's found under the kitchen sink and can normally be identified by the fact that it's covered in dust, paint and streaks of limescale and hasn't been operated since the Bay City Rollers were last in town.

If your mains stop tap isn't under the sink, try removing the plinths from the kitchen units and seeing if you can spot either the tap itself or the pipe coming up from the floor into the house; in old houses this could be a lead pipe, in newer houses it could be a black or blue plastic pipe. If that doesn't work, try looking next to the toilet in the downstairs cloakroom. If it's not there hunt around the house, bearing in mind that it ought to be close to an external wall and no higher than 150mm from the ground. If your home has been extended at some point the tap may still be next to the original external wall, which could now be in the middle of the house.

Turning off the mains stop tap will turn off all the water in your home ... sooner or later. Depending on what hot- and cold-water systems have been installed in your home, it will turn off most – if not all – of your cold water straight away, and possibly all of your hot water. Those hot- and cold-water taps that are still running will be drawing their water from storage tanks, and with the mains tap turned off these will empty, and the taps stop, after about ten minutes. Closing this stop tap will even ensure that your central-heating system eventually empties. In other words, if you turn this tap off before you go on holiday and your home does spring a

> ⚠ **NOTE**
>
> Throughout this book you'll notice that I mix metric and imperial measurements. For example I might describe something as "a ¾" tap connector going into 15mm pipe". Rightly or wrongly, this is the generally accepted industry nomenclature at the moment so we might as well use it here. Some might look at this and suggest that the plumbing industry embraced metric with all the enthusiasm that you and I might embrace a skunk. And you know what? I'm not sure I'd argue with that suggestion.

leak, the leak will stop before it has totally destroyed your home – although you shouldn't just rely on this tap.

So, the question is: having found your mains stop tap, does it work? If the answer is no then you have to repair it as soon as possible, and we'll be going through the steps to do this in 'The mains stop tap' (Chapter 7). If it is working, then you might want to put a label on it so that everyone else in the house knows what it is. Whilst you're at it, make sure the tap is easily accessible and, most importantly, use it whenever you're planning on leaving the house empty for any length of time.

Some people labour under the illusion that you have to have the cold water on if you want to run your central heating. Fortunately you don't, so in winter, leave the CH ticking over at 5°C and turn the mains stop tap off if you're going away for any length of time.

Having found the mains stop tap we'll now venture outdoors and find the next most important tap – the water company's 'outdoor stop tap'.

> **TIP** *If someone comes to your door claiming to be from the 'Water Board' or 'the Water Authority', don't let them in. The businesses in charge of delivering water to your home are now all called 'water companies', a fact of which bogus callers are often unaware, hence their use of the wrong nomenclature. Regardless of the name, though, if they don't have official ID don't let them in anyway.*

Where is the outdoor stopcock?

In an ideal world the mains stop tap will always work. In the real world this tap will wait until your kitchen is under two feet of water, and then break. Fortunately the people who design these things are, without exception, utter and complete pessimists. As such your water system has been designed with two stop taps: one in your house, the other outside.

The latter tap is owned by your local water company and is generally found just on the other side of your property line. As such it's usually set into the pavement just opposite your house, hidden underground beneath a little round plastic cover. I hope you've noted the words 'generally' and 'usually' here, because in reality this tap can be just about anywhere – I once found one on the corner of a country lane, hidden under sheep poo, about three miles away from the house we were working in. The cover the tap is hidden under also changes depending upon its age; new ones are round plastic covers, older ones tend to be rectangular and metal. Occasionally the covers will have something useful written on them like 'water', but don't bank on this.

If the weather is fine you can turn 'find my stop tap' into fun for all the family. However, if you prefer an easy life ring your local water company, give them your details and they'll get back to you with a diagram showing you where the tap's located. Having found it you need to lift the cover to get at it:

1 Run a flat-headed screwdriver around the edge of the lid to loosen off the dirt.

2 If it has a round plastic cover you can push the screwdriver into one of the grooves on the edge and lift it up.

If you have an older rectangular cover you'll see that one side is hinged and opposite this is a groove. Insert the screwdriver into this groove and lift. The lid might be stiff and need a little persuasion with a claw hammer, but it should come up eventually.

The third type of cover you may come up against is plastic and round but resists your attempts to lift it with a screwdriver, or pretty much anything else. This is because it's a screw lid, and the only way you can tell this is because the lid has two large grooves in it. I assume there's a proper tool to open these, but if there is it seems to be a closely guarded secret known only to an inner circle of water company engineers. The best approach I've found is to jam a large, blunt screwdriver into one groove and then tap it vigorously anticlockwise with a hammer.

3 What you should find once this cover is up is a foam or polystyrene protective cover and beneath this a pristine plastic tap with, or without, a water meter. Sadly, this isn't always the case; sometimes the hole is completely filled with murky water, occasionally it's filled with soil with no sign of a tap whatsoever.

Bearing in mind that this tap is the property of the water company and that it's there to allow the water to be turned off in an emergency, you have the right to demand that the water company maintain this tap in good working order. As such, if you can't get at this tap or if it doesn't readily turn, ring your water company and ask them to repair or replace it – they'll usually tell you they'll get around to it in about ten working days, but in my experience they've usually sorted it all out in just a day or so.

4 If you have one of the modern stop taps you can turn it off by inserting the plastic 'key' into the hexagonal tap head and turning it clockwise – the key is often already in the tap head. Beware of turning the key too hard; it's only plastic, and it's not unknown for the hexagonal head to shear away, leaving you without any water until the tap is fixed.

If you have one of the older taps, think about asking for it to be updated. If you'd rather not, go and buy yourself a stopcock key. This is just a metal rod with a bend on one end. You place this over the tap and gently turn it clockwise to turn the water off.

The latest version of the outdoor stop tap has a round lid which when lifted will reveal a foam cover and under this a pristine ¼ turn tap. The tap itself is a stumpy bar of white plastic which can be turned by hand or with the old fashioned stop tap key. When open the head points towards the meter - or the place where the meter can be fitted. When closed the head is at right angles to the meter. Just to make things more confusing sometimes this key acts like a standard tap and needs to be turned clockwise until it can't be turned any further. Life eh!

So there we are; regardless of what sort of system you have in your house you can now turn the water off. This might not stop the water flow immediately, but it will stop it eventually. Depending on how the water and heating services were installed in your home, there will be a number of other taps and valves that will let you stop the flow of water more quickly and we'll look at these in the next Section.

Identifying hot and cold water systems

Your home is usually equipped with a number of taps and valves to help you close down and drain your hot and cold water pipes. What type and where they are varies depending upon what type of hot and cold water system was installed in the first place. So we'll start off by trying to figure out how your home has been plumbed.

As a general rule there are two types of domestic water systems: stored or mains-fed.

Stored systems

In a stored, or low-pressure, system the cold-water mains coming into your home is used to fill large storage tanks situated as high up in your home as possible – usually in the loft. The water that comes out of your taps is fed from these storage tanks, with the

result that turning off the mains cold water doesn't instantly stop water coming out of your taps. In fact, if the tanks are particularly large it can take 15–30 minutes for them to empty and for the flow from your taps to cease. In an emergency this can be a particularly anxious and invariably damp 30 minutes, so if your home does use stored water you need to find out where the valves are that will turn this water off immediately.

Mains-fed systems

In this scenario (also known as a high-pressure system), all cold water comes directly from the mains supply, so switching off the cold-water stop tap will immediately, or almost immediately, stop the cold-water taps running. This mains cold water is also used to 'power' your hot water, meaning that it's the pressure of the cold water that's used to push the hot water out of your taps. As such, turning off the mains stop tap deprives the hot water of pressure and the hot-water taps stop running as well. Bear in mind that you might still have a vessel where the hot water is stored, but without the cold-water mains pressure to push hot

water around, it can't reach the taps and just sits in this hot-water cylinder where it can do little or no harm – theoretically.

Just to make matters a little bit more complicated, most houses will have a mix of stored and

mains-fed water. In a typical circa 1960s house, for example, the cold water will be mains-fed to the kitchen then delivered via storage tanks to the bathroom, whilst all of the hot water will be from a stored supply.

There are many advantages to a high-pressure system but the important one for us here is that turning off your mains stop tap will immediately turn off all the hot and cold water in your home. The only disadvantage is that an awful lot of water can leak out whilst you're trying to find your mains stop tap – unless you've already found it, tested it and told everyone else in the house where it is.

What cold-water system do you have in your home?

So how do you tell what you have in your home? Well, for the cold water the easiest way is to follow the flow chart below:

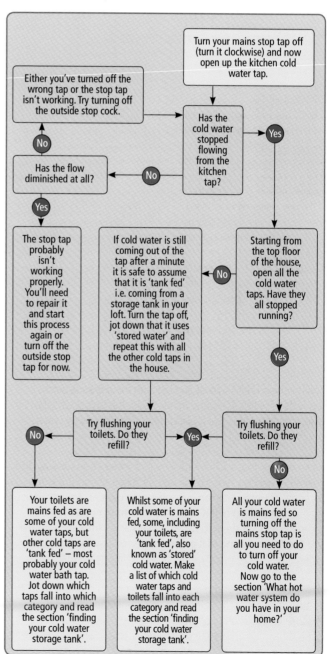

What hot-water system do you have in your home?

If you've just finished going through the cold-water flow chart the mains stop tap should still be turned off. If it isn't, start this section by turning it off. Now close the cold-water tap over the kitchen sink and open the hot-water tap.

The great thing with the hot water in your home is that it tends to work the same for all the taps, so if you can identify the system for one tap you've identified them all.

So now that you've opened one hot tap all you need to do is sit back and see what happens. Generally speaking you'll see one of two things:

- The hot water stops flowing almost immediately – see 'High-pressure hot-water systems' on page 19.
- The hot water carries on running and running (usually for three to ten minutes) – see 'Low-pressure hot-water systems' below.

LOW-PRESSURE HOT-WATER SYSTEMS

If the hot water is still flowing after a minute then what you have is 'stored hot water', also known as 'low-pressure hot water' or 'gravity-fed hot water'.

In this system you store cold water in your loft, or as high in the house as is practicable. This cold-water store is linked by a single pipe to the hot-water cylinder below and the whole system works using gravity: when you open a hot-water tap, the cold water in the storage tank runs down the connecting pipe and into the base of the hot-water cylinder, pushing the hot water out of the top of the cylinder and off towards the tap.

When you turn the cold mains tap off you stop the cold-water storage tank from being refilled, but the hot water itself will not stop running until the tank is empty. This tank usually holds at least 114 litres (25 gallons) but can be much, much larger. As such it's not a good idea to rely on emptying the cold-water storage tank every time you want to stop your hot water running.

Note that it's the cold-water storage tank that needs to empty to stop the hot water running, and not the hot-water cylinder below it. As was mentioned earlier, it's the cold-water tank that provides the 'power' to push the hot water out. So once the cold-water tank is empty you're left with a full hot-water cylinder but no power to push the water out. At this point the hot water stops flowing. I only mention this because some people think that if the hot water has stopped flowing the hot-water cylinder must be empty. This is a wholly incorrect assumption that can get you very wet indeed.

Bearing all this in mind, what we really want to do is just break the connection between the cold-water storage tank and the hot-water cylinder, and fortunately there's a tap provided for just such an occasion.

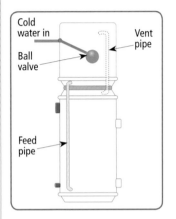

NOTE

There are two oddities that ought to be mentioned here. The first is the 'Fortic cylinder', which was very popular in flats where there was no loft to put a storage tank. In essence it's a standard cylinder with the storage tank stuck directly on top. To make this more difficult to see they then cover the entire thing in foam.

The harder oddity to identify is called a 'Primatic cylinder'. In this arrangement the storage for both the central heating and the hot water is provided by a single large tank in the loft, and the hot water is kept apart from the central-heating water by nothing more than a bubble of air inside the cylinder. God knows how this ever came to be regarded as a good idea, but it was. The only sure way to identify this system is to find the label on the cylinder that says 'Primatic', although the fact that you only have a single storage tank in the loft is a clue.

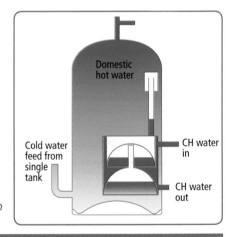

In the diagram (left) you can see a 'gate valve' shown on the connecting pipe. As a general rule this is in the airing cupboard alongside the hot-water cylinder, usually just a little higher than the top of the cylinder. Turning this off will cause the hot water to stop running almost immediately.

Sadly there are invariably a number of valves and taps in the airing cupboard, but the one we're looking for comes down from the cold-water storage tank above and runs into the very base of the cylinder.

More often than not the tap itself is either round and red or a chrome valve with a coloured 'lever' handle.

Having found this tap, turn it clockwise to close it and check that this has turned off your hot water completely. If it does, label it 'Hot-water stop tap' and

The new lever valves

The old round red valves

let everyone else in your home know where it is. If this tap doesn't work read 'The hot-water stop tap' (Chapter 7).

Finally, if you have gravity-fed hot water it pays to know where your cold-water storage tank actually is, and to know that it's in good condition, so jump to 'Finding your cold-water storage tank' on page 21.

HIGH-PRESSURE HOT-WATER SYSTEMS

If turning off your mains stop tap turned off your hot water as well as your cold, then your hot water is known as a 'high-pressure hot-water system' and is delivered either instantaneously or via an unvented hot-water cylinder. In both cases the hot water is 'powered' by the cold-water pressure; so when you turn off the mains stop tap you deprive your hot water of this pressure and as a result nothing comes out of the hot-water taps.

> ⚠ **WARNING**
>
> In larger houses with umpteen bathrooms the high-pressure hot water is boosted by a device called an accumulator. This is basically a store of pressurised air that keeps the hot water flowing at high pressure even when there's a large demand. The only reason for mentioning this here is that the accumulator can also cause the hot water to keep flowing even when the cold water has been turned off. The clues that may make you suspect your home has an accumulator are that the hot water acts as if it's from a stored water source – ie it keeps flowing minutes after the cold water has been turned off, yet you can't find a store of cold water in the loft – and that your hot water seems to come out of the taps at high pressure. Another clue is that your house is so big you have to dig out a map to find your way around.

The easiest way to determine if you have an unvented cylinder or are using instantaneous hot water is to hunt around your house looking for an unvented cylinder. This is almost impossible to miss on account of it being a very large, steel-lined cylinder with numerous pipes leading in and out of it – see 'Identifying an unvented cylinder' (page 11).

However, just to make things a bit more difficult, this cylinder can be pretty much anywhere in your home, from the basement to

the loft, so start off looking in the obvious places – in the airing cupboard upstairs or next to your boiler – and then widen your search.

Unvented hot-water cylinder

If you manage to find this cylinder then the next step is to find the tap that allows you to immediately turn off your hot water, and only your hot water. The tap in question will be on the cold-water mains pipe running into the cylinder, and in an ideal world it will have a label saying something along the lines of 'Cold-water mains'. If you live in a less than ideal world, and most of us do, you should still be able to identify this valve by its position:

1 There will be a pipe going into the base of the cylinder.

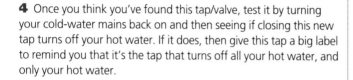

2 Follow this up and you'll usually come to a complex array of pipes and valves called a 'combination valve'.

3 There are a number of pipes leaving this combination valve but generally only one will have a tap on it that can be turned off.

4 Once you think you've found this tap/valve, test it by turning your cold-water mains back on and then seeing if closing this new tap turns off your hot water. If it does, then give this tap a big label to remind you that it's the tap that turns off all your hot water, and only your hot water.

Before we move on it's an idea to note two facts about unvented cylinders:

- They don't need a cold-water storage tank up in the loft – although your central heating system might.
- They can be dangerous if not serviced regularly by a suitably qualified plumber (G3 unvented cylinder certificate).

Combi boiler

If you can't find a cylinder in your home then you almost certainly have an instantaneous hot-water system, which might be delivered via an unvented hot-water heater but is most probably delivered via a 'combi boiler'.

In this system there's no stored hot water anywhere in your home. Instead the boiler will detect when you open a hot-water tap and immediately start heating up some mains cold water to supply this tap.

The upside of

this approach is that you only heat up water when it's needed, rather than heating up a store of it that might just go cold again without ever being used. The downside of this approach is that generally it can't provide very much hot water. So if you live in a one-bedroom flat, get yourself a combi boiler, and if you live in a five-bedroom mansion with numerous ensuites, don't.

TIP *For some reason many people get confused with the terms 'combi boiler' and 'condensing boiler'. A combi boiler delivers both central heating and instantaneous hot water, whereas a condensing boiler is just a very efficient type of boiler. As such a combi boiler can be a 'condensing combi boiler', while a condensing boiler doesn't have to be a 'combi'.*

If you have a combi boiler you'll need to identify the cold-water mains pipe coming into it. The best way to find this pipe is to find the manual for your boiler and read it.

Having found the pipe, look for the valve that allows you to turn off the cold-water mains to the boiler. Closing this valve won't affect your central-heating system but will turn off just your hot water.

THERMAL STORE

One other system that you might happen across is called a thermal store, which is basically an ordinary cylinder (vented or unvented) working in reverse. Instead of the cylinder containing the water that comes out of the taps and the coil inside containing your central heating water, the store of water in the cylinder is heated directly by the boiler and it's the coil that contains the water that emerges from your taps. As cold water passes through this coil it picks up the heat from the water in the cylinder and emerges from the coil as high-pressure hot water. In this instance the hot water will stop flowing as soon as you turn off the cold-water mains.

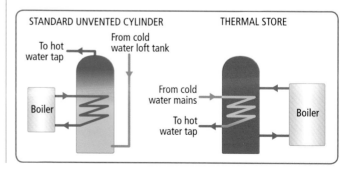

STANDARD UNVENTED CYLINDER   THERMAL STORE

Finding your cold-water storage tank

If all your hot and cold water is high pressure then you shouldn't have a cold-water storage tank in your home. I've used the word 'shouldn't' rather than 'won't' for the simple reason that strange

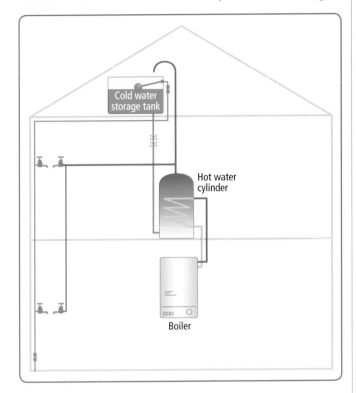

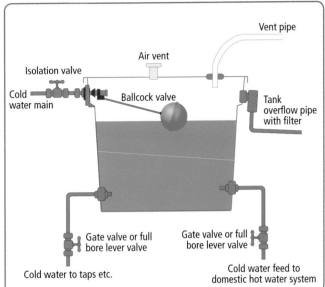

- At the very top there should be a bent pipe coming down into the tank; this is the vent pipe coming from your hot-water cylinder.
- Near the top of the tank you'll see the ball valve and the mains cold-water pipe running into the back of it.
- Slightly lower down will be a pipe running gently downwards towards the eves of the house. This is usually in plastic and will terminate outside, just under your guttering. This is the overflow pipe, and it's this pipe that will save your home from flooding if the ball valve should ever fail, so be nice to it!
- Towards the bottom of the tank, and sometimes underneath it, you'll find the pipes feeding hot and cold water to your home.

things do happen in the plumbing world, so you might want to have a read of this section anyway.

For now these are the only pipes we're interested in, but whilst you're up in the loft you might want to take note of the following:

If your home uses stored hot or cold water then it delivers this water to your taps via the wonder of gravity. As such any search for the storage tank should start as high up in your home as possible, usually your loft – if it's not in your loft try the top of the airing cupboard.

Once up in the loft you need to be aware of the fact that there may be more than one storage tank up here. In a typical 1970s layout your loft would contain a small tank feeding the central-heating system and, usually close by, a much larger tank feeding your hot-water cylinder and some, if not most, of your cold-water taps.

Hopefully you've now found one or more storage tanks and, by comparing the sizes, deduced which is the cold water and which is the central heating. If the size difference between your storage tanks isn't too clear try taking a look inside them. 'Oh my God!' is a phrase commonly used to describe the inside of the smaller central-heating tank, whereas the cold-water storage tank *should* be a clean, clear reservoir of fine water. Another identifier is the fact that the smaller CH tank should only contain about 10–15cm of water, whereas the cold-water storage tank should be almost full.

Hopefully your cold-water storage tank looks a bit like the diagram shown top right. Sadly, the close-fitting lid with an air-vent in it and a rubber seal around the vent pipe is more often than not replaced with a slab of polystyrene, some old carpets, bits of cardboard or, somewhat ironically, umpteen back-issues of *Ideal Home*. However, you should still be able to use the diagram to identify the pipes leading into and out of the tank.

- Is the tank made from plastic or is it galvanised steel? If it's steel think about replacing it immediately, and in the meantime don't touch it. I'm serious! Galvanised tanks haven't been fitted to houses for eons so it's almost certainly on its way out. It might last two minutes, it might last another five years, but in the interim just breathing heavily nearby might be enough to cause it to disintegrate. See 'Fitting a new cold-water storage tank' (Chapter 8, page 108).
- Is/are the tank(s) insulated? If not, you need to read the section on fitting a Byelaw 30 kit ('Insulating your storage tanks', Chapter 5, page 48).
- What does the ball valve look like? If it's looking a bit knackered read the section on repairing/replacing a ball valve ('Broken ball valve', Chapter 8, page 98).
- Does the cold-water mains pipe leading up to the ball valve have an isolation valve just before it? If not it might be an idea to fit one – see 'Isolation valves' (Chapter 7, page 79).
- Does the overflow pipe look in good condition? Is it blocked at all?
- Have a look inside the tank. Is it clean? Would you want to drink this water? If the answer is no, read 'Insulating your storage tanks' (Chapter 5), where we talk about fitting a 'Byelaw 30 kit' – an essential collection of bits'n'bobs designed to keep the water in this tank in pristine condition.

The odds are that the pipes coming from the cold-water storage tank aren't labelled so you're going to have to work out which are which:

1 As already mentioned, if you follow one of these pipes you should find yourself standing next to the hot-water cylinder and in front of you will be the tap to turn off this cylinder. You can now test and label this tap.

2 More often than not there's only one other pipe coming from

the storage tank and the tap to turn this off is usually right next to the tank. This pipe will deliver stored cold water to the house. Check that the tap works and give it a label that makes sense to you. If the tap doesn't work have a look at 'The hot-water stop tap' (Chapter 7, page 78) – same tap, just different water temperature.

3 In some houses there will be other pipes leading from the base of the cold-water storage tank. These usually deliver cold water to specific showers. Follow them to figure out which shower they go to, then find the tap that turns the water off and test and label it.

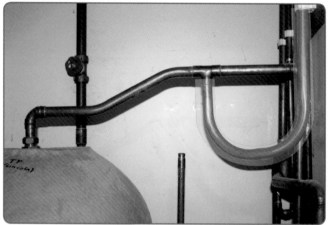

4 For each cold-water pipe feeding a shower there should be a corresponding hot-water pipe. These should start from the very top of the hot-water cylinder and have a tap so that you can turn off the hot water to just this shower. Find which one works which shower, test the tap and label everything.

Going on holiday

If you have high-pressure hot and cold water to your house then just turn off the mains cold-water tap before you go away. Having done this, open up the kitchen sink taps to drain the system and close them again before you lock up. This only takes five minutes and could save you an absolute fortune.

If you have stored hot and cold water turn off the cold-water mains tap as well as the appropriate hot and cold feed taps before going away. I wouldn't advise draining down a stored system because it can result in an airlock, which can be a nightmare to clear. Yes, by not draining down the tanks and pipework there's still a risk of a leak, but it's now unlikely to be a flood.

What type of central heating do you have?

Whilst there are alternative approaches, most homes now use a 'wet' central-heating system to keep the whole house nice and warm. These all involve a boiler or a heat pump of some description that's used to heat up water. Once hot this water is circulated around the home, transferring its heat to either traditional radiators or underfloor pipework, which in turn heat up the rooms in which they're situated.

When all is working this is a great system, but what we're interested in here is what happens when it *isn't* working, or, more to the point, what do you do when all that hot water suddenly makes a bid for freedom, escapes from its pipework and starts flooding your home?

In order to answer this question you first need to know what type of CH system you have. Fortunately this is a fairly easy task on account of there only being two types: gravity-fed systems and high-pressure systems.

Gravity-fed central heating

In a gravity-fed system water rushes into your home via the stop tap and, with barely a pause for thought, races up into your loft. Once there it passes through a ball valve and starts filling a small tank, called the feed and expansion tank. Good old gravity now kicks in and under its benign influence water flows out of this tank, down into your boiler and into all your radiators – hence the term 'gravity-fed'.

Once the system is full the ball valve closes and that's pretty much it; the water within your boiler and radiators now circulates around and around, possibly for years at a time, never leaving the CH circuit. This is the reason why you can turn off your cold-water mains when you go on holiday, yet still leave your central heating on to keep your home warm. It's also the reason why turning off

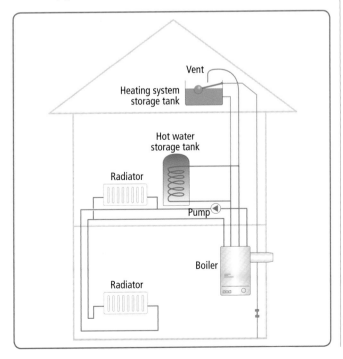

your mains stop tap will not immediately stop water pouring out of a leaking central-heating pipe.

So how do you find out if this is the sort of CH system you have? Well, the easiest way is to pop back up to your loft and look for this 'feed and expansion' tank. It will be a fairly small tank (roughly 500mm long by about 300mm wide by 300mm tall) and

if there are two or more tanks up in your loft this one will be the smallest. If you take a look inside you'll see that the tank is less than half full and that the water it contains looks foul enough to put you off the idea of drinking it, which is just as well because it looks better than it tastes – apparently!

If you can't find this tank in your loft, try looking in the airing cupboard. In terms of location you just need to bear in mind that it must be the highest point in your whole CH system, so it has to be higher than any of your radiators, ideally quite a bit higher.

If you can't find this tank you probably have a high-pressure CH system. If this is the case, jump to the next section.

TURNING OFF THE WATER TO YOUR GRAVITY-FED SYSTEM

When you remove water from a gravity-fed CH system – either deliberately or because you have a leak – the water level in the feed and expansion tank (F&E tank) drops. This causes the ball valve to open and fresh water pours back into the tank to replace the water that was lost.

So to drain the CH system you could just turn off the cold-water mains tap. This is a nice and easy approach and in an emergency is by far and away the best thing to do. Sadly, turning off the cold-water mains turns off the cold water to the entire house. This generally isn't a problem for half an hour or so but can be a right pain in the derriere if you need to work on the CH for the whole day or can't find a plumber to fix the leak until after the Christmas holidays.

An alternative is to stop the ball valve from opening as the water level in the tank drops:

1 Lay a piece of wood over the top of the tank.

2 Loop some twine under the arm of the ball valve and around the wood. Remember, all you're trying to do is stop the arm of the ball valve from dropping. The ball valve is not trying to escape so there's no need to encase it in twine and there's no need to yank the arm up into the air, which could permanently damage the valve.

(diagram labels) Vent · Heating system storage tank · Hot water storage tank · Radiator · Pump · Boiler · Radiator

3 As the system drains and the tank empties check your wood and twine to make sure it's all stable and secure – if the twine breaks or the wood falls into the tank you're going to get pretty wet, pretty quickly.

Because it only takes one slip or knock to upset the 'wood and twine' method you may want to take a different approach and stop water actually reaching the ball valve – and only the ball valve – in the first place.

All you need for this is a tap, and if your luck is in you'll actually find one by just tracing the cold-water pipe back from the ball valve. In a perfect world, not only will you find this tap but it will actually work!

To test it, turn the head clockwise to close it and then push down the arm of the ball valve. If the tap works nothing will come out of the ball valve. If the tap is broken then water will continue to emerge from the ball valve whilst you have the arm depressed.

A slightly more modern approach to the red gate valve shown above is the isolation valve.

If you can't find a tap anywhere on this pipe you might want to consider putting one in – see 'Fitting an isolation valve' (Chapter 7). The best place for the isolation valve is just behind the ball valve itself; in fact you can buy isolation valves that fit directly on to the back of the ball valve.

NOTE

Some of you may have looked at the diagram on the previous page and thought, 'Why on earth don't I just put a tap on the pipe leading from the base of the tank to my CH system?'

On the face of it this seems a fairly sound idea ... but it isn't. You may recall that this tank is called the 'feed and expansion tank' (F&E tank), and it's called this because it has three distinct roles. The most obvious ones are to provide the water pressure for the CH system and to act as a reservoir, feeding more water into the system as and when required, eg when you bleed a radiator.

The third and far less obvious role of this tank is to give the water in your CH system somewhere to expand into when it gets hot.

When you heat water it expands by about 4%. As a rule of thumb the average domestic CH system contains about a 100 litres of water, so, when heated, it expands by about four litres – a little over a gallon. In a gravity-fed system this expanded water moves up the CH feed pipe and into the F&E tank. This is why, when cold, the feed and expansion tank should be about half empty, so that the water can expand into it without overflowing the tank and running out of the overflow pipe.

But if you put a tap on this feed pipe and accidentally close it, the hot water can no longer move up into the tank. If something can't expand its pressure will rise instead, and high-pressure hot water can be very dangerous.

In fairness there's another safety device – the vent pipe – to stop this happening, but it's never a good idea to rely on only one safety device, so never, ever put a tap on the feed pipe leading from the F&E tank.

Once you've isolated the F&E tank, either by tying up the ball valve or fitting an isolation valve, you can start to drain down your CH system, which is discussed later.

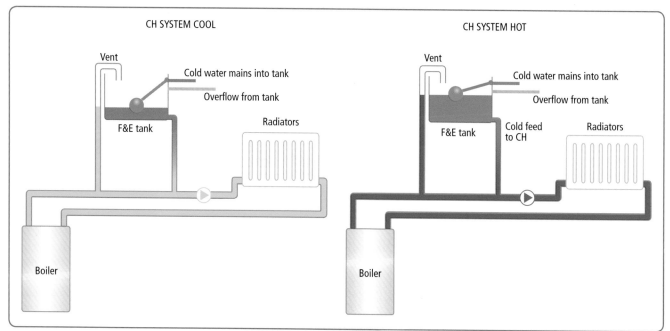

CH SYSTEM COOL

Vent

Cold water mains into tank

Overflow from tank

F&E tank

Radiators

Boiler

CH SYSTEM HOT

Vent

Cold water mains into tank

Overflow from tank

F&E tank

Cold feed to CH

Radiators

Boiler

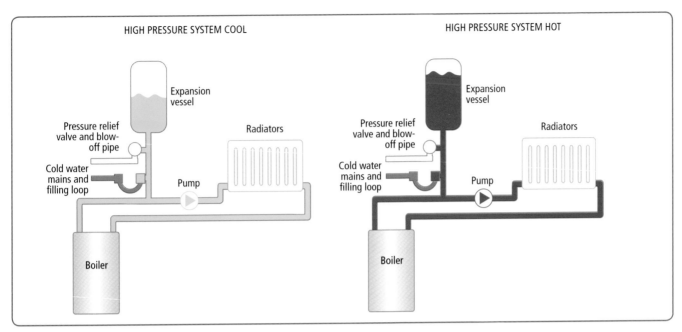

HIGH PRESSURE SYSTEM COOL

HIGH PRESSURE SYSTEM HOT

Expansion vessel

Pressure relief valve and blow-off pipe

Radiators

Cold water mains and filling loop

Pump

Boiler

High-pressure central heating

Most new CH and boiler installations are high-pressure systems because, on the whole, they're both better and cheaper to install. All high-pressure systems will have the following components:

- **A pressure gauge** – This can sometimes be confused with a temperature gauge, so check the scale. It should be in bar or psi.

- **An expansion vessel** – This is usually bright red but can also be grey. Note that your hot-water system could also be pressurised, so make sure you don't confuse the hot-water expansion vessel with that of the central heating. The hot-water expansion vessel is usually white or blue and is fitted very close to the hot-water cylinder itself.

- **A filling loop** – Sadly the design of the filling loop varies enormously. The one shown is usually external to the boiler.

You'd think that these critical components would all look the same from manufacturer to

manufacturer but they don't. However, the gauge is pretty standard to look at and it's always within view of the filling loop, since you have to use one whilst looking at the other to set the pressure correctly. In turn, both of these components are often built into the boiler or, if they aren't with the boiler, they're often very close to an external expansion vessel, which, being large and red, is often the first component you'll find.

So the best way to check for a high-pressure system is as follows:

- Check the boiler.
  If it has a pressure gauge built into it you definitely have a high-pressure system. This gauge can be behind a cover or just under the boiler, so have a good look around and maybe dig out the boiler manual to see if it says where the pressure gauge is located.
- If the boiler doesn't have a gauge, search the loft and airing cupboard for an F&E tank. If you find one you don't have a high-pressure system. If you don't find one you *probably* have a high-pressure system, but it's not a certainty.

TIP *If you do find an F&E tank check that it has water in it. When CH systems are converted from gravity-fed to high pressure the F&E tank is disconnected from the CH system but is often still left in the loft.*

- If you find nothing close to the boiler try looking in the airing cupboard or wherever the hot-water cylinder is located. When external to the boiler the filling loop is usually as per the one shown to the left. It's always close to a pressure gauge and a bright red expansion vessel is invariably close by. However, remember that the hot-water cylinder could have its own expansion vessel.

- Read the boiler manual and/or ring the manufacturer and ask them if your boiler is a combi boiler. If they say yes, then you have a high-pressure CH system.
- If the manufacturer says it's not a combi boiler ask if it's a 'system boiler', and if it isn't can it be converted to a high-pressure system? If the answer to both questions is no then you definitely don't have a high-pressure system.
- If any doubts remain call a local plumber and ask them to check for you. It should take a qualified plumber less than half an hour to do this check.

So, having established that you have a high pressure CH system, how does it work? Well, in the gravity-fed system the pipework is filled from a tank in the loft; but in a high-pressure system the tank is dispensed with and the system fills directly from the mains cold water.

To do this there has to be a mechanism that allows the CH system to be kept completely separate from the rest of the water in the home whilst it's running, but that also allows the CH and the cold-water pipework to occasionally be linked together so that the cold-water mains can 'top up' or refill the central-heating pipework. The gadget that does all this is called a 'filling loop'.

USING A FILLING LOOP

As already mentioned, the filling loop can vary from manufacturer to manufacturer. However, the principle is always the same, and nearly all filling loops that are external to the boiler work as described below:

1 A filling loop is simply two taps joined together by a bit of flexible hose. One tap is connected to the cold-water mains and the other is connected to the CH system.

It's always best to fill the system when it's cold, so turn the boiler off and give the system time to cool.

2 In normal operation both taps should be closed and the hose disconnected. In reality the hose is invariably left connected to both taps, so you can usually jump step 3.

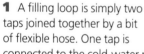

3 To fill the CH system you first connect the flexible hose to both taps. You should only need to hand-tighten the wing nuts.

4 The taps themselves are either small plastic levers or metal slots. The former you turn by hand for the latter you need to insert a screwdriver into the slots to turn them. In both cases when the tap or slot is aligned with the tap body and hose the valve is open. When they cut

across the tap body the valve is closed.

First open the tap on the cold water side – usually connected to the thinner pipework. This will fill the hose with water, so check for drips. If you find one just give the hose wing nuts a tweak with a set of pump pliers.

5 Close to the filling loop there should be a pressure gauge. When cold most CH systems need to have a pressure

of between 1 and 1.5bar. If the pressure gauge shows a value lower than this you need to top it up. If the pressure is higher than this you need to lower it. The easiest way to do this is to bleed a radiator – see 'Bleeding a radiator' (Chapter 12, page 173). When hot the pressure should be about 2bar.

6 So the cold tap is open and you can see the current pressure. Now open the tap on the CH side. You'll immediately hear water racing into the system and you'll see the pressure gauge start to rise. Keep the tap open until the pressure reaches about 1bar. However, note that most combi boilers seem to prefer the pressure to be closer to 1.5bar. (If in doubt ring the manufacturer, tell them the model of your boiler and ask for the correct pressure when the system is cold.)

7 Once the pressure is correct close the CH tap. Now close the tap on the cold water side and disconnect the hose. Leave the hose connected to the cold-water tap so that you don't lose it.

As mentioned, some manufacturers seem to go out of their way to create their own version of the filling loop.

Below is a filling loop as supplied by Worcester Bosch. In this case the flexible hose is replaced by a white 'key'. This is pushed

into the filling loop assembly and turned to lock it in place. Once you've done that you turn the white knob – shown to the left of the key – controlling the cold water.

Below is Ideal Boilers' version. This works as per a standard filling loop, but the flexible hose is replaced by a rigid copper tube and the taps look a little different.

I could fill the rest of this book with alternative approaches to

the standard filling loop, so you're best off digging out your boiler manual, finding the section on filling the system and reading all about the filling loop provided for your particular boiler.

There were two reasons we wanted to find the filling loop. Firstly, it proves that you have a high-pressure system, and secondly, it's the one device on your CH system that you'll want to use on a fairly regular basis, as every time you bleed a radiator you'll need to reset the pressure.

However, for this chapter the main thing we're trying to do is just confirm what type of central-heating system you have. The pressure gauge and filling loop has proved that you have a high-pressure system. So, aside from draining down, what do you do if you have a leak in this system?

Hopefully you should already know what the answer to this question is, because it's 'Nothing!' For gravity-fed central heating you first have to isolate the cold water from the central-heating pipework, but for high-pressure systems that's already been done – you only ever connect the cold water and the CH system together when you need to repressurise the central heating.

Draining down your central heating

If you have a high-pressure CH system you don't have to worry about isolating the CH from the mains cold-water supply. If you have a gravity-fed system you'll have to complete this task before you can start draining down the system – see page 23.

If you've been blessed by fate, a leak in your CH system will be relatively easy to deal with because you'll find a working drain cock near a door, just waiting to drain down everything before too much damage is done. Sadly, most of us have a fairly ambiguous arrangement with fate so, whilst we'll start off with the notion of a working drain cock, we also need to look at the alternatives.

USING A DRAIN COCK

1 A drain cock starts off life as a small brass tap with a square head that can be turned with a small spanner. It usually ends up as a large, misshapen, blob of paint, but even then the original 'tap-ness' of it can still be vaguely discerned.

To look for it, start off by checking the radiator in the hallway, near the front door. It will usually be on the pipework very close to this radiator. If you can't find anything, check out the other radiators downstairs – the kitchen one is another favourite.

An alternative to a separate drain cock is a radiator valve with a drain point built into it. The design of these varies from manufacturer to manufacturer but they all have a short protruding spout, usually ribbed so that a hosepipe can be attached to it.

2 Having found your drain cock you need to place something underneath it to stop any drips landing on your carpet. If yours is a brand new or expensive carpet you might even

want to consider rolling the carpet back from the radiator before opening the valve, otherwise try one of more of the following: a sponge, a shallow tray, or a sheet of plastic with a dust sheet over it.

3 If you have one, fit a jubilee clip to your hosepipe and then fit the hose to the drain cock. If you're right in the middle of an emergency and can't find a jubilee clip, just attach the hose directly to the drain cock, put a sponge under it, and make sure there are no kinks in the hose.

4 Run the hosepipe outside, ideally into a drain but on to the garden if need be – just bear in mind that the water is usually very dirty.

5 Now open up the drain cock. Usually this means turning a little square head with a small spanner. As you turn the head it slowly unscrews itself from the body of the valve and you should then start to hear water emerging from the drain cock

and racing away down the hosepipe.

Some of the latest drain cocks use an Allen key to open them and this approach is often used with drain cocks attached directly to radiator valves. If this is the case, just find the right Allen key and use this to open the valve.

6 Once you have the water running out of the hose, check for leaks around the drain cock to make sure that the water is running away from the hose safely.

7 You've now started draining the system and in most instances this will cause any leak to stop completely. However, the system will still contain a lot of water. To drain the system completely wait until the flow of water from the hosepipe has started to ebb and then carefully open up the bleed valve on each radiator, starting with those at the top of the house and working your way down.

To do this you need to insert a bleed key into the radiator and turn the bleed valve anticlockwise. If the system is almost empty you'll

TIP *A drain cock is just a tiny tap that's fitted to a CH system and then forgotten. Because it's rarely used, the rubber washer within it has a habit of corroding, and this can cause two effects that you need to be aware of:*

Firstly you open the drain cock and nothing happens. This is because the rubber washer is stuck to the base of the drain cock and is therefore still blocking the opening. The easiest way to get out of this pickle is to tighten up the valve as tight as you can and then try again. By squashing the rubber washer you can usually either cause it to disintegrate or dislodge itself from the tap valve body – either way the valve opens up, which is the essential point in an emergency.

The second effect leads on from the first in that the washer is often damaged when you open the drain cock. This doesn't make itself known until you refill the system and suddenly find water leaking everywhere. To avoid this always unscrew the drain cock completely once the system is drained and just check that the washer is intact and in good condition.

There are new drain cocks coming to market that use a spherical steel ball rather than a washer. The advantage of these is that there's nothing to corrode, so they should always open up when needed and not leak afterwards.

Ground floor radiators are often fed by pipework dropping down to them from the first floor. If this is the case these radiators will not empty unless they're on the same leg of pipework as the drain cock you've just opened, *ie* if the pipework comes down the wall from upstairs and feeds just one radiator then the only way to drain this radiator would be to open up the drain cock underneath it. If there isn't one you'll need to take the radiator off to drain this section of pipework – see *draining the central heating via a radiator*.

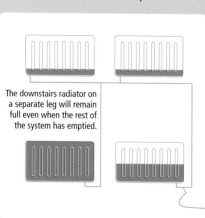

The downstairs radiator on a separate leg will remain full even when the rest of the system has emptied.

If you have a leak on the ground floor pipework and there isn't a drain cock on that leg of pipework it's still a good idea to drain the system using whatever drain cock is available. It mightn't stop the leak completely but it will slow it down an awful lot as the pressure of water in the pipe is reduced.

hear air being sucked into the radiator at this point. If it's still full water will come out of the radiator. If this happens close the bleed valve and wait a while longer before trying again.

DRAINING THE CENTRAL HEATING VIA A RADIATOR
If you can't find a drain cock on your CH system you're going to have to drain the system from a radiator point instead. This takes an awful lot longer to complete, so once you've completed it and the CH is completely drained think long and hard about fitting a drain cock to the system.

1 To drain from a radiator you first need to take the radiator off the wall – see 'Removing a radiator' (Chapter 12).

2 With the radiator removed, you now have to connect a hose to the radiator valve. The easiest way to do this is to push a piece of copper tube into the end of your hosepipe and keep it there using a jubilee clip.

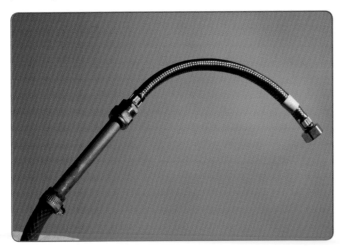

3 Now push a flexible tap connector on to this copper tube – a push-fit connector is easiest. Note that radiator valves come in two sizes. Most modern ones are ½in and most old ones are ¾in. With this in mind you'll need either a 15mm to ½in flexi tap connector or a 15mm to ¾in flexi tap connector. And yes, you need to use this mix of millimetres and inches if you expect anyone at the plumbing merchants to understand you!

4 Now attach the tap connector to the radiator valve. Tighten it by hand and then give it an extra three-quarter turn using an adjustable spanner.

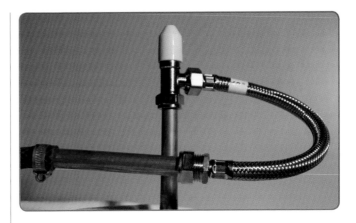

Attach the flexi hose to the radiator tail.

5 Now run the hose to an outside drain point and open the radiator valve by turning the head anticlockwise. You should hear water starting to race away at this point so check for leaks at the radiator and then check everything is OK outside. If you don't have a hosepipe or can't connect it to your radiator valve you'll have to put a container under the radiator valve, open the valve, fill the container, close the valve, empty the container and repeat *ad nauseam*.

Bearing in mind that you'll often be doing all this in an emergency you should be able to see by now why it's a good idea to check for drain cocks on your CH system and, if you haven't got any, get some.

Refilling your central heating
Firstly make sure you've checked the washer and closed the drain cock that you used to drain the system in the first place ... then check it again. If you took a radiator off the wall, make sure you've put it back on again. Once you're sure all is ready you can refill the system.

If you have a high-pressure CH system you'll need to refill it using the filling loop as described earlier.

If you have a gravity-fed system you'll have isolated the feed and expansion tank from the mains cold water before you started this job. If this is the case it's just a matter of reversing the process you used in the first place, *ie* open the isolation valve or remove the string and wood.

Now you'll need to bleed all the radiators – see Chapter 12.

Words of advice
I hope you read this chapter whilst resting your feet upon the sofa and calmly sipping a cup of tea. However, I also hope you've pondered on just how prepared you are if an emergency did happen. Do you have a hosepipe ready to hand? Do you have a drain cock on your CH system? If so, what are the odds of it working, *ie* is it a nice piece of untarnished, shiny brass or is it caked in 25 years' worth of paint and as likely to work as a lettuce-powered laptop?

I wouldn't advise draining down a CH system just for the sheer hell of it all, but it might pay dividends to just fit a hose, open up the drain cock and see if it starts draining. If you do this with a high-pressure CH system you'll need to top up the water again via the filling loop. If you have a gravity-fed system, don't worry – it'll fill up again all by itself.

Home Plumbing Manual

3 PLUMBING WEAR, TOOLS AND MATERIALS

Essential safety equipment   31

Planning   32

Essential tools   32

Having a go at your own plumbing can save time and money as well as be very rewarding ... provided you do it right. Do it wrong and you could end up wet, massively out of pocket and rapidly running out of excuses when your partner tactfully asks 'What went wrong?'

There are two important steps you can take to ensure that it all goes swimmingly. The first is to make sure you're using the right safety equipment. Not only does this ensure that you're still in the land of the living when the plaudits are being handed out, but it also means that you'll still be in good enough shape to complete the lap of honour for a job well done. The second step is to make sure that you have the right tools for the job.

Essential safety equipment

KNEE PADS

The two most common plumbing injuries are sore knees and aching elbows. The first comes from kneeling down on the floor for hours at a time without using knee pads or some other form of cushioning.

External knee pads have the advantage of fitting over any trousers, or even shorts, and will save your knees. On the down side they almost always rub at the back of your knees, and I personally find this too much irritation to bear. However, they're relatively cheap, they do the job and you can buy them at most plumbing merchants or DIY stores.

Gardening pads have the advantage of being a nice comfortable way of saving your knees. The obvious disadvantage is that they are not attached to your knees, so you'll find yourself constantly having to tug them around after you. On the plus side they're comfortable, they're the cheapest option, and most DIY and gardening centres will sell them.

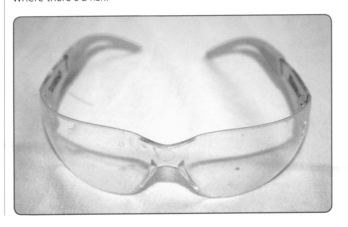

What the professionals use are work trousers with little pockets in the knees that allow foam rubber pads to be inserted into them. The advantages are that they're comfortable and are always there when you need them. The disadvantage is that the trousers themselves can be quite expensive and the pads do sometimes fall out – unless you go for the top-loading option. These can be ordered online or purchased from your local plumbing merchant or DIY store.

SAFETY BOOTS

If you do enough plumbing or general DIY you will, sooner or later, drop something on your feet. Not only can this be extremely painful but it can leave you unfit to do anything other than operate the TV remote control for weeks on end. If a month of mid-morning TV sends a cold shiver down your spine – and let's face it, it ought to – then invest in a pair of steel toe-capped boots. There's a vast range available so you ought to find something that will suit your feet, your pocket and your desire for sartorial elegance.

EAR DEFENDERS

Most plumbing work doesn't involve the prolonged use of power tools but if this ever looks likely get a pair of ear defenders. These come as either little memory foam earplugs or huge great headphones. I've consistently failed to follow this very sage advice and as a result now have tinnitus. It's a right pain in the ear, so please be more sensible than me.

SAFETY GLASSES

Again, most plumbing work doesn't put your eyes in any great danger, but it's best to have a pair around for those odd jobs where there's a risk.

DUST MASK

If you're going to spend any length of time in a loft you really want to put one of these on first, and the same applies if you're channelling out walls or floors with a power tool.

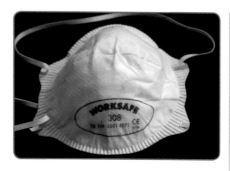

Fortunately, for most plumbing work the cheap little white paper masks are more than adequate, but if you're likely to spend a lot of time in a dusty environment go for the masks with a cartridge in them, ideally a replaceable cartridge.

 WARNING

A standard dust mask won't protect you from asbestos, as the fibres are so small that they'll just pass straight through the mask and into your lungs. If you suspect something in your home might be made of asbestos, then if it's in good condition just leave it well alone, or if you have to go near it consult an expert first.

PPE PACK

A lot of plumbing and building merchants offer what they refer to as a 'PPE pack' (personal protective equipment), which is a pair of safety glasses, a dust mask and a set of ear defenders. This is a cheap and easy way of ensuring that you always have these things to hand.

Planning

This is the best safety aid ever invented but is often the one least used. Never just jump into a job; always sit down and plan what you're going to do, what tools you'll need and what the possible dangers are going to be. Having recognised the dangers think about how best to avoid them, and only then think about starting the job.

Yes, this is all astonishingly obvious and, yes, many of you will be accusing me of being patronising. Yet every weekend the local A&E department runneth over with people who didn't take this advice but wish to God they had.

"I told you it was a stupid idea to try to fix the disposal with your tie on."

COMMON SENSE

Along with planning this is the other essential piece of safety equipment that's left behind when people start a job. The two most important aspects of common sense are admitting to yourself that a job is just a bit too much to take on right now, and not overdoing it at the start.

Plumbing is not overly strenuous. However, if you're new to DIY you'll find yourself using muscles that don't normally get exercised. The classic case is when someone suddenly decides to take up a screwdriver and then spends the next four hours fitting screws. The odds are that the next day their elbow will be absolutely killing them. Why? Because they've just managed to develop tennis elbow without the long and tedious process of learning to play tennis. The big problem with tennis elbow is not the immediate pain, though that's bad enough – it's that once you have it, it takes an age to go away. So don't get it in the first place. Buy an electric screwdriver and pace yourself; plumbing is not a race!

Essential tools

Most of the basic tools in plumbing are tools you'll probably have around the house already and, if you haven't, they're well worth buying. For more elaborate work you'll need to judge for yourself if it's worth buying the necessary tools for a one-off job or if it's going to be cheaper to just call a plumber.

ADJUSTABLE SPANNER

It's a good idea to have at least two of these and it pays to buy a good pair. Cheap spanners tend to loosen off very readily, just when you don't want them too,

and the jaws don't open wide enough to accommodate all the pipework you'll encounter. Adjustable spanners can be used on chrome nuts and, provided they don't slip, shouldn't cause any damage.

PUMP PLIERS

If you can't get at a nut with your adjustable spanner then these will be just what you need. They're also ideal for larger joints.

There's a right and a wrong way to use pump pliers. If you have them the wrong way around, as you apply pressure you'll find the handles try to open up in your hand. If this is the case just turn them over and try again – this time you should find that the handles actually start to tighten the more pressure you apply.

Whilst pliers are good for most things don't use them on chrome or anything ornate, as the jaws have ridges on them that will ruin any chrome finish.

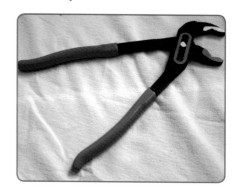

TAPE MEASURE

It's hard to think of a DIY event that doesn't at some stage require the use of a tape measure, and plumbing is no exception.

SCREWDRIVERS

You'll generally only use one or two flat-headed screwdrivers, small and medium, and usually only one Pozi or Phillips, the medium one.

Mind you, at some point you'll need another sort so it's a good idea to buy a cheap set to start off with and then buy good quality replacements for the ones you use most often.

Cordless powered screwdrivers can save a lot of time and elbow wear and are well worth buying. For larger screws an impact driver is an absolute godsend, and personally I wouldn't be without one.

UTILITY KNIFE

These are horribly dangerous but enormously useful and at some point or another you'll wish you had one.

HAMMER DRILL

At some stage or other you'll need to drill a hole, and if you're going to drill it into masonry you'll either need a lot of time on your hands or you'll need a hammer drill. The price of drills can vary enormously. Cordless hammer drills are much more expensive than those

with a flex but are a damn sight more convenient to use. SDS drills are more expensive than ordinary drills, but then again they don't have a chuck key that can get lost at a critical moment. Often the difference in price is due to the power of the drill, the torque it can generate, the rate at which it can hammer away and how durable it is. As a general rule plumbing, especially DIY plumbing, doesn't require a top of the range drill, so just buy one that suits your pocket.

Alongside the drill you'll need some drill bits. The most useful ones are 6mm and 8mm, but if you're going to drill into tiles you'll need a decent tile drill, and if you're going to hang a washbasin off a wall you'll usually need a 10mm or 14mm bit.

BASIN SPANNER

Taps are a piece of cake to fit and remove ... provided you can get at them. Sadly, you usually encounter them after some numpty has fitted them to a washbasin, bath or sink and as a result they're nigh on impossible to get at with traditional spanners or pliers. Fortunately, they invented basin spanners for just this occasion.

To use them you need to twist the head to that it's at roughly right angles to the shaft. If you twist it one way the spring-loaded jaws will tighten as you turn the shaft clockwise

(tightening the nut); if you twist the head the other way they'll tighten as you turn the shaft anticlockwise (loosening the nut).

To get the jaws onto the nut just push gently and the jaws will open up to accept the nut.

Now just turn the shaft and they'll tighten on the nut. If they open up you've got the head the wrong way around, so turn it right over and try again.

BOX SPANNERS

For washbasin and bath taps a ½in and ¾in box spanner can be very handy. For kitchen sink and most monobloc taps (see 'Fitting new monobloc taps' in Chapter 10) a set of three box spanners

is essential to get them on and off.

The sets will usually fit into each other to get at the more difficult kitchen taps, and to use them you can either put a spanner on the end or push a screwdriver through one of the holes in the base.

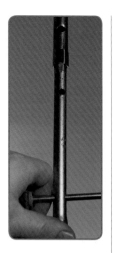

1. Turn knob to open jaws and push onto pipe
2. Rotate around pipe
3. Tighten knob a bit more

PIPE CUTTERS

You can just hacksaw through pipework, but this invariably results in a slightly angled cut, with a sharp frayed edge and sometimes a slight distortion to the end of the pipe – all of which can cause the subsequent joint to leak. As such it's better to use a proper pipe cutter as these will give you a perfectly angled cut with a nicely rounded edge every time. There are many varieties to choose from so let's look at the options:

Round pipe cutters

These are ideal for cutting into copper pipework in situ. To operate them just slip them over the pipe in question and then turn them in the direction indicated by the arrow. How much you'll need to turn them depends on the age of the pipework – older copper is usually much thicker – and the age of your pipe cutters, but just persevere and you'll be through.

The upside of these cutters is that their shape allows you to get them on to most pipework, regardless of location. The downside is that they can require a fair bit of strength to use, especially if you're cutting larger diameter pipework. You can get around this to an extent by turning them with a pair of pump pliers.

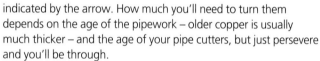

Another downside is that they only fit one size of pipe, so you may need to buy three of them: 15mm, 22mm and 28mm.

Handled cutters

With these you turn the knob anticlockwise to open up the jaws, push them over the copper pipe and then turn the knob clockwise until you feel a bit of resistance. Now rotate the entire cutter around the pipe, turn the knob clockwise a bit more and rotate the cutter again. Repeat this until you're through the pipe.

The advantage of these cutters is that little or no strength is required because the handle gives you all the leverage you'll need, and you're controlling how deeply each turn cuts into the pipe. Another plus is that one pipe cutter can be used for most, if not all, the different pipe sizes you'll come across. On the downside it's rare that they can be used on pipework that's in situ because you

just don't have enough room to turn them. Another downside is that you can distort the pipe if you try to cut through too quickly, *ie* you're tightening them too much after each turn.

Cutting waste pipework

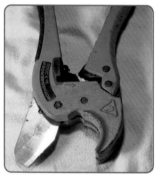

The cutter right is the same as the round copper pipe cutters but sized for the plastic waste pipework (32 and 40mm) that you come across in the home.

The cutters below are ideal for cutting all sizes of plastic pipework, from 8mm all the way through to 40mm. With the smaller diameter pipes you can just cut straight through, but for the larger sizes (32 and 40mm) it's best to tighten the jaws on the pipework and then rotate the cutter so that you slowly slice through the pipe – but be careful, as they do have a habit of cutting a spiral if the pipe isn't perfectly positioned. Personally, I rotate them once to start the cut and then squeeze the handle to slice through.

PUSH-FIT STOP-ENDS

Before you cut through any pipe you will of course have made sure that the pipe in question is empty ... and as sure as eggs are eggs, one day you'll make a mistake and discover that the pipe is still full. At this point you can either put your thumb over the end of the pipe and hope you have the strength to stop the flow – nigh on impossible if it's under mains pressure – or race screaming down the stairs to find your stopcock while your home gently floods. The problem with putting your thumb over the pipe is that you're now stuck until someone comes to your rescue and either turns the water off or brings you a large bucket.

To avoid these disaster scenarios, never cut into any pipe without first making sure that you have at least two push-fit stop-ends that fit the pipe size in question and are sitting within easy reach. If the worse does happen, quickly finish the cut and then push the stop-end on to the leaking pipe. Usually you'll discover that the water gushing out is just the few litres left in the pipework after you've drained the system, but it's best to make this discovery in a relaxed fashion after you've stopped the leak and got yourself a bucket or sponge to soak up whatever water is going to emerge.

HACKSAW

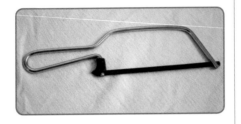

There are always some pipes that can't be cut with a proper cutter and at this point the hacksaw comes to your aid. If you do use one for cutting pipework try your best to keep the cut at right angles to the pipe, and once you're through clean the edges with a bit of abrasive strip or a small flat file.

ABRASIVE STRIPS

However you choose to join your pipework together you'll always need to make sure the two ends are cleaned first. In the old days we used wire wool for this but this often resulted in shredded fingers, so we ditched it and started buying abrasive strips instead. It's more expensive than wire wool but it's so much safer and easier to use.

PTFE TAPE

Possibly the most over used item in plumbing is PTFE tape (or polytetrafluoroethylene tape, as my partner in plumbing perversely insists on calling it, mainly for the perplexed looks this creates down at the plumbing merchants).

In most instances the waterproof seal in plumbing is created by a washer or an olive. However, there are occasions (radiator valve tails being a classic example) where a 'jointing compound' is applied to a threaded joint. The compound fills the hollows in the thread and creates a waterproof seal as the thread is subsequently tightened into the joint. The most popular of these jointing compounds is PTFE tape, mainly because it's simple to use and leaves no messy residues.

Sadly PTFE tape is also used to replace damaged washers, where it rarely works for more than a few months, or is often just wrapped around pipework in the vain hope that this will stop a leak.

A much-overlooked fact about PTFE tape is that there's a right and a wrong way of using it.

The right way is that you hold the reel of tape in your right hand, take a bit of the tape off the bottom of the reel and apply it to the thread, which you're holding in your left hand. It's important that you take the tape off the bottom of the reel – just try taking it off the top and you'll soon see why.

Now wrap it clockwise around the thread, *ie* in the same direction that you'll subsequently tighten the thread into the joint.

The reason for this is that you want the tape to tighten up and work its way deeper into the thread as you tighten the joint. If you wrap it the other way the tape actually loosens off as you tighten the thread. As a result you end up with a thick ridge of PTFE tape at the end of the thread, which not only looks messy but also means there's little or no PTFE inside the joint to prevent the water from getting out.

> ### ⚠ WARNING
>
> Some immersion heaters are deliberately designed with an anticlockwise thread to stop you putting the wrong one into the cylinder. If you're going to use PTFE in this instance be sure to wrap it around the thread *anticlockwise*.

BOX OF WASHERS

Most of the leaks that you'll come across in the home are caused by a washer finally giving up the ghost, so it's always handy to keep a box containing a variety of fibre and rubber washers around the home. Most DIY and plumbing stores will sell a 'variety box' that will cover 90% of occasions. A box of tap washers is also useful to have around. We'll be going through how to re-washer a tap in the coming chapters, so you'd best get a box of washers beforehand.

WATER VACUUM

Most DIY stores sell these things and they're absolutely wonderful. I've already intimated that there's nothing worse than cutting into a pipe only to find that it's still full of about three litres of water. In the bad old days this would leave you racing around looking for bigger sponges, buckets and towels. However, with a vacuum cleaner that can suck up water as well as dust you can just flick a switch and watch the water being sucked away, giving you plenty

of time to prepare a stop-end ... just in case. They're also great for draining away all the water prior to soldering some pipework, and they can make changing an old radiator a relatively humdrum experience, regardless of the colour and quality of your carpeting.

4 DEALING WITH EMERGENCIES

What to do in an emergency   37

Causes of leaks   37

Burst pipes   38

Frozen pipes   42

What to do in an emergency

I've no idea what odds William Hill will give you, but I'd advise against popping in and betting on a leak-free life. Rather, bet that you'll get a leak and that when it happens Sod's Law will be in full force; the leak won't occur at a convenient time, it will have issued no warnings, and you'll have just fitted a lovely new beige carpet.

Hopefully the leak, when it happens, will be a minor incident, but sadly this isn't always the case. Over the years I've come across a number of homes where the staircase was fit only for white-water rafting, the living room had been turned into a ducks' paradise and the loft hatch was doing a fair impression of Niagara Falls.

It's not just the enormous hole in your wallet that results from these disasters; it's the heartbreak as you stand there, stunned into silence, while a lifetime of possessions floats by. What makes it worse is that, whilst you might not have been able to stop it happening altogether, you could easily have turned this deluge into a relatively nondescript dribble.

So, having accepted that at some time or other you're going to get wet, what can you do to minimise the damage and reduce the odds of it happening in the first place?

Well, as already discussed, you need to know where your mains stop tap is and, having discovered its whereabouts, how and when to use it. By now you should also know what type of hot- and cold-water systems you have in your home, what kind of CH system and, most importantly of all, how to quickly drain them in an emergency.

Of course, it's always best to not get the leak in the first place, and whilst it's almost impossible to totally achieve this goal you can come pretty close.

Causes of leaks

Eventually *all* pipework will start to leak, but it's rare to find a leak caused purely by the pipework corroding away. Far more common causes are:

FROZEN PIPEWORK

At some point during the winter we'll get a call to repair a leak caused by water freezing in an uninsulated pipe. Virtually every time the leak will be caused by the homeowners going on holiday and deciding to turn off their central heating for the duration.

Let's clarify something here; you can turn off your hot and cold water whilst leaving your central-heating system running. In fact this is exactly what you should do every time you leave your home empty during the winter months.

Of course, you'll always find someone who disagrees with this: 'Ah, yes,' they say, 'but it costs money to keep my house warm when I'm away, and anyway, I've already insulated all the pipework in my loft.'

Well, firstly you only need to keep the boiler ticking over: a mere 5°C is enough to stop anything freezing and that's not going to break even the flimsiest of banks. The second point is that it doesn't really matter if you've insulated all the pipework in your loft if you then let the temperature inside your home, where none of the pipework is insulated, fall below freezing.

So will keeping the central heating on solve everything? Sadly, no. In most houses the central heating will only keep 90% of your home warm. Outside of this 'heat envelope' the temperature can drop below freezing regardless of your central-heating system and in these areas you need to insulate the pipework. The most obvious

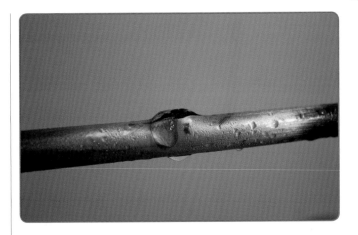

area for pipe insulation is your loft, but porches, garages and conservatories also need to be checked.

In the next chapter we'll be going through the steps needed to insulate your pipes and tanks but for now we just need to bear in mind one other important issue: just because you insulated your pipework last year doesn't mean it's still insulated now! It's a sad fact that to a rodent your brand new foam insulated pipework looks like a very long, very tasty and bizarrely moreish baguette. In fact one of the easiest ways of telling if you have a rodent problem in your loft is to check the state of the pipework up there; if you can see lots of shiny copper close to small heaps of neatly nibbled foam then you have guests.

If this is the case you need to consider either removing the guests or changing your insulation material – hessian insulation and foil-wrapped glass fibre insulation are the rodent equivalent of sprouts, so whilst they may still take a nibble they're unlikely to finish the meal.

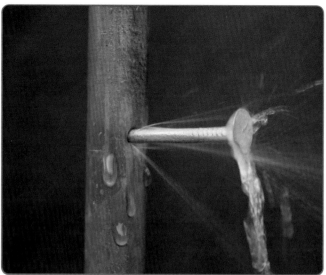

BALL VALVES FAILING

Most homes in the UK will have at least three ball valves: one in the toilet cistern, one in the cold-water storage tank in the loft and one in the smaller feed and expansion tank, which is also in the loft. The ball valve is a great invention, but eventually all ball valves will fail. To make matters worse, ball valves are not, as a general rule, failsafe. In fact they usually fail downright dangerously, releasing a huge amount of water into a relatively small tank or cistern. To cope with this, every tank or cistern has an overflow pipe.

The theory is that when the ball valve fails all the excess water races out of this pipe and is deposited safely outside of the house. Meanwhile, back in reality, it has taken ten years for the ball valve to fail, during which time everyone has forgotten all about it. The tank it feeds is covered with dust and carpets and the occasional piece of polystyrene. The overflow pipe leading from the tank is now filled with spiders' webs, dead leaves and other detritus that is best not examined too closely. During the years both the plastic tank and the plastic overflow pipe have expanded and contracted as the weather rages around the house and as a result the pipe is now connected to the tank by nothing more than grit and determination.

Eventually the great day arrives and the ball valve fails. The water level quickly rises in the tank until it reaches the overflow pipe. We all hold our breath as the water starts to run down the overflow pipe only to be met by a wall of debris, and once again the water level in the tank starts to rise ... and rise, until the tank overflows into the house. And this is the main problem: the overflow often fails at much the same time as the ball valve.

So make a note of the following:

- Check the state of your various ball valves once a year. It only takes two minutes and can be part of a general 'spring check' for the whole house.
- If you opt to change the ball valve (see Chapter 8) always check the state of the overflow pipe and clean or change this at the same time.
- Better still, fit a Byelaw 30 kit – also known as a Regulation 16 kit – which, by happy coincidence, we'll be discussing in Chapter 5.

DIY

DIY must be the most common cause of household floods. It's not that there's anything inherently risky about DIY; it's just that many folk hammer first and ask questions later, the most common question being 'Where did all that water suddenly come from?' So try to keep the following basics in mind when working about the house:

- Most pipework is routed just under the floorboards and in some cases 'just' really is the operative word. So before you nail through a floorboard try lifting it first to check what's underneath.
- If this isn't possible only use short nails/screws. The floorboard is usually only about 19mm thick so pick a nail or screw that's not going to go straight through it and into the pipework below. Also, consider using glue, either instead of or alongside smaller nails/screws. The glues in question usually feature the word 'nails' – or the lack thereof – in their names and really do work.
- Yes, it does make sense to use metal detectors, but remember that they don't always work and they won't find plastic pipework, which is the material of choice for new builds.
- Pipework also runs down walls. Bear this in mind if you're hanging a painting, especially if you're hanging it near a radiator.
- If you are unfortunate enough to drive a nail or screw through a pipe, try to fight the desire to immediately pull it back out again. This isn't always an easy thing to do but the fact remains that this errant screw or nail is plugging up most of the hole, so leave it be whilst you drain off the water.
- Before you start a DIY project make sure you know how to turn off and drain all the water in your home.

Burst pipes

The most common emergency, or certainly the one most likely to get the heart racing, is a burst pipe. If the pipe has burst because you've just put a nail through it DON'T REMOVE THE NAIL!

Odds are that by the time you've read that statement you'll have removed the nail – it's an instinctive reaction, but sadly the wrong one, as the dribble of water will have now turned into a fountain.

To get things quickly under control follow these steps:

1 Open all your taps. This mightn't stop the leak straight away, but if the leak is in a hot or cold-water pipe it will reduce the flow rate.

2 Close the cold-water mains stop tap.

3 Having read the previous chapters you should now know how to turn off any stored hot and cold water in your home, so go do it.

4 If the leak was in a hot or cold-water tap it should have stopped by now. If it hasn't, the odds are you've got a leak in the central-heating pipework. This can usually be confirmed by the colour and smell of the water – CH water has a slightly chemical smell and is often black with gunk. To deal with this you need to drain your CH system as described in 'Draining down your central heating' (Chapter 2).

Once you have the water under control again you can take a deep breath and decide how best to repair the leak. The simplest options are:

LEAK REPAIR TAPES & COMPOUNDS
The advantage of repair tapes are that you don't need to cut into the pipe – in fact you can often effect a temporary repair whilst the pipe is still leaking, although it's a good idea to at least reduce the pressure first before wrapping the tape around the

pipe. Sadly, it's rarely that easy; the pipework is often under floorboards and usually running alongside at least one other pipe, so wrapping the pipe can be a bit of a nightmare. The other downside to all this is that the repair isn't permanent.

CLAMP REPAIR KIT
The advantage of this approach is that you don't need to cut the pipe and you can, in theory at least, apply it whilst the pipework is still under pressure – although you might wish to don a pair of swimming trunks first. A further advantage is that the clamp is relatively easy to use when there's other pipework close by. However, as with tape, the repair shouldn't be considered a permanent solution, although it should tide you over for a few days whilst you get your breath back.

The rest of the repair approaches listed below are permanent but all require that you cut out the bit of pipe where the leak is. With this in mind you'll first need to ensure that the pipework has been drained, and you'll also have to buy a pipe cutter. Yes, you can get through the pipe with a hacksaw, but the cut is often at an angle and the edge is usually sharp and distorted. All of which is bad news if you want to ensure a permanent repair, so it pays to buy a proper pipe cutter – see Chapter 3 *Essential tools*.

PUSH-FIT REPAIR KIT

Tools and materials – for copper pipe
■ Push-fit repair kit
■ Tape measure
■ Pipe cutters
■ Abrasive strips

Tools and materials – for plastic pipe
■ Push-fit repair kit
■ Tape measure
■ Pipe cutters
■ Pipe strengtheners

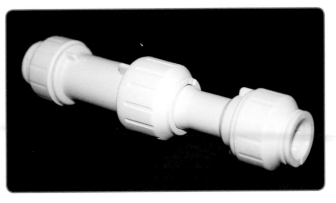

The advantage of this approach is that it's quick and easy to fit, the repair is permanent and it works particularly well if the leak occurred in plastic pipework. The kit is also designed to slip over the pipe, which is very, very handy if there is absolutely no movement in the pipework.

On the downside, if you have copper pipe in your home you'll need to maintain the earth continuity by bridging the plastic repair with an earth bond.

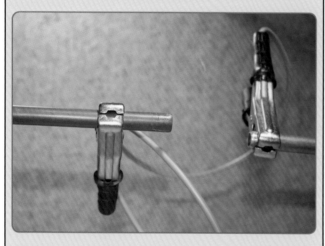
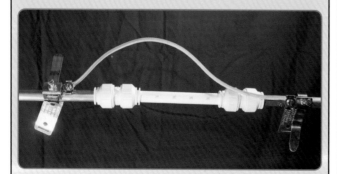
It would be a good idea to read 'Plastic push fit pipework' (Chapter 6) before following these instructions.

1 To use this kit you need to drain the pipework and identify what size it is (usually 15mm or 22mm – if in doubt buy both kits and return the one you don't use).

2 Now cut out the area of pipe containing the leak, making sure that you're left with perfectly round and undamaged ends. The maximum and minimum distances you can cut out are usually in the instructions that come with the kit.

With plastic pipe you'll get a much cleaner and straighter cut if you purchase some plastic pipe cutters.

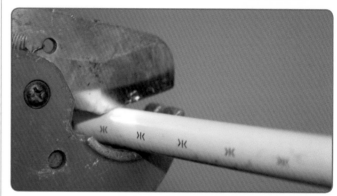

3 For copper pipework, once you've cut it use an abrasive strip to clean the two cut ends. Make absolutely sure that the pipe ends are clean and perfectly round.

For plastic pipe you'll need to insert a plastic strengthener into each pipe end. These ensure that the plastic pipe maintains its size and shape under pressure and are essential whenever you fit any joint on to plastic pipework.

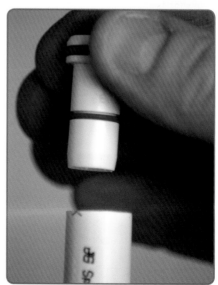

NOTE

If the leak was caused by the pipe freezing, the area around the leak is often warped by the ice pressure. The pipe often looks perfectly round but it's actually a little wider now than it should be and you'll struggle to get a watertight fitting on it. With this in mind it's best to cut the pipe about 10cm on either side of the actual leak. If need be you can use two repair kits and connect them together with a length of new pipe – or just use two push-fit straight fittings and a length of new pipe in between.

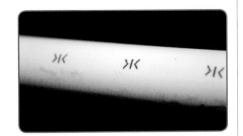

4 To ensure that the push-fit fitting is pushed on correctly plastic pipe is marked with little chevrons. Note where the end of your pipe is in relation to these markers and make sure this gap is the same after you've pushed the fitting on.

For copper pipe make a mark about 30mm from the end of the pipe for 15mm pipework and about 38mm for the larger 22mm pipework. When the fitting is fully pushed on this mark should only just be visible.

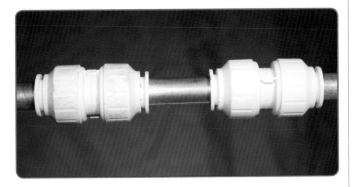

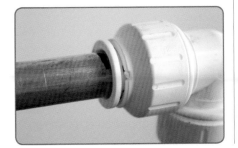

5 Now dismantle the fitting as per the instructions, dampen the pipe ends and the fittings with a little water or washing-up liquid to let the fitting slip on easier and push the fitting on to the pipework. If it's gone fully in you should see that the end of the joint is resting very close to the pencil marks you just drew; if it's still a fair distance away you need to push a little more. Note that one end of the fitting will slip fully on to the pipe – this is useful, but it makes the pencil marks you've added doubly important, as when you slip the fitting back again you need to make sure you haven't slipped it *too* far back.

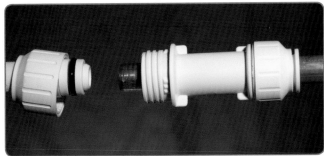

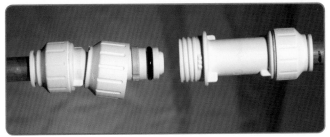

6 Fit both ends of the fitting on to the pipework.

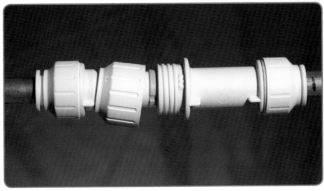

7 Then push the two ends together.

8 Now tighten the fitting.

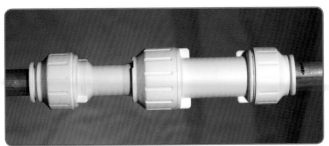

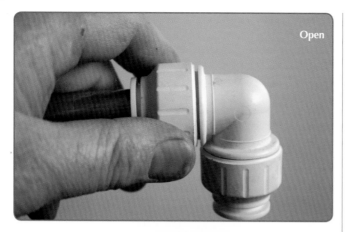

Open

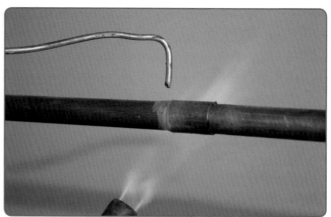

9 To finish, you need to hand-tighten all the joints to stop them being pulled off again by accident. You can test this by trying to pull the fitting off. You shouldn't succeed!

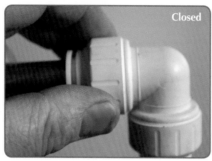

Closed

10 The most common problem with push-fit is that the joints aren't pushed in far enough or are pushed in at such an angle that the rubber O-ring inside them gets pulled out of position instead of sliding over the pipe to form a watertight seal. With this in mind it pays to get someone else to turn the water back on whilst you keep an eye on your new repair. If water starts to come out again ... yell!

11 If you made your repair on copper pipework you should now fit two earth bonding strips on either side of your repair to maintain earth continuity.

COMPRESSION REPAIR KIT

This is far more awkward to use but creates a permanent fix and maintains earth continuity. You'll need some give in the pipework to fit this kit, but by unscrewing the heads and slipping these and their associated olives on to the pipework first you can get away with very little give. There's a technique to using compression joints so read 'Using a compression fitting' (Chapter 6) before you try this approach.

SLIP-SOCKETS

These are the professional approach to repairing a leak and have the advantage of being permanent, maintaining earth continuity, and not requiring any give in the pipework. Sadly, they also require a fair amount of skill with a blowtorch or electric pipe heater, so are best avoided unless you're happy soldering joints – see 'Using integral solder ring fittings' or 'Using end-feed fittings' in Chapter 6. However, if you're feeling confident you just need one ordinary socket and one slip-socket.

The slip-socket, as its name suggests, will slip completely on to

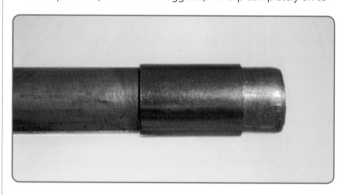

the pipework. As such you can cut your new section of copper to size, slide the slip-socket on to it, put one end of the new section into your ordinary socket, then put the new section in place and push the slip-socket into position. It's usually a good idea to mark a line on the pipework so that you can see when the slip-socket is in the right place.

Frozen pipes

If your hot or cold-water pipework freezes the giveaway sign is that all or some of your taps suddenly stop working. If you have a gravity hot-water system (see Chapter 2) then pop into your loft and check out your cold-water storage tank. It's not unknown for the ball valve to have frozen shut even if all of the pipework is insulated. If this is the case it's pretty obvious – the tank is almost dry and the 'ball' is hanging in midair. If it's not the ball valve check the pipework in the loft to see if any sections are uninsulated – it only takes a small section to freeze to stop the flow completely. Before you attempt to thaw any pipework, first check that no sections have burst. Whilst everything is frozen the burst pipe is filled with ice and no harm is being done, but once you start thawing it all out this will change ... for the worse.

end of the flue. This water runs back along the flue and down into the boiler before draining away via a little plastic pipe. These days this plastic pipe tends to be run into an indoor drain, but this isn't always possible and, instead, it'll run out through an external wall and go into an outdoor drain. Once outdoors this pipe is subject to freezing. If it does then the condensate water starts to rise, first filling the pipe and then filling the boiler. Most boilers at this stage take fright and automatically shut themselves down, which is usually when you first notice that the condensate pipe is frozen.

To thaw it out just follow the earlier guidance – ie get out your hairdryer, or failing that pour a kettle of hot water over the pipe. Once you're sure the pipework is ice-free, insulate the exposed pipework (see Chapter 5), press the reset button on your boiler and away you go.

So, you're happy that there's no burst pipework and reasonably sure where the frozen section is. It's now time to break out the hairdryer or, if you haven't got a hairdryer, an electric paint-stripper. What you don't want to be using is a blowtorch or anything else that issues flames. Not because it won't work – it will – but because you might end up burning down your home at the same time, for which running water is small consolation.

With all the defunct taps open, remove any insulating material and start playing the hairdryer along the pipework, or over the ball valve if you think this is the problem. Have someone down below to keep an eye on the taps and get them to shout up as soon as water starts to emerge.

Once you have the water running again, double-check for leaks and then insulate everything (see Chapter 5) before descending from the loft.

If the frozen pipework is outdoors you can use hot water from a kettle to thaw it if the lead on your hairdryer isn't long enough.

FROZEN CONDENSATE PIPEWORK

New boilers are terribly efficient, but this particular silver lining comes with its very own cloud.

To be efficient new boilers extract as much heat as they can from the flue gases, so much heat in fact that some of this gas condenses back into water before it even leaves the

 WARNING

Don't be tempted to just disconnect the condensate pipework and let the condensate drip into a bucket. The most obvious reason for this is that the bucket might overflow, but on some boilers it's also possible for flue gases to now escape into the room.

5 INSULATING PIPEWORK AND STORAGE TANKS

Insulating your pipework 45

Applying foam insulation to pipework 46

Insulating your storage tanks 48

Fitting a Regulation 16 kit (Byelaw 30 kit) 49

Insulating your pipework

In many older homes the loft area will contain hot- and cold-water pipework, possibly some CH pipework, and probably one or more water storage tanks. Because heat rises another thing a loft can contain is an awful lot of heat, heat that's been paid for but which is now wilfully wasting away up at the top of the house. To prevent this huge waste of money most of us have gone out and purchased thick swathes of insulation, which is a good thing. However, every silver lining has a cloud, and the cumulonimbus of loft insulation is that the loft itself can now become very, very cold.

During a prolonged cold spell the loft can become so cold that the pipework within it, or rather the water in that pipework, starts to freeze, and as it freezes this icy water expands. Ordinarily this is no big deal, but when this water's in a pipe the room available for expansion is severely limited. Sadly, the water doesn't take this very well and eventually it causes the pipe to rupture and split.

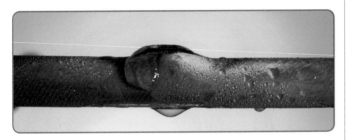

At this point you could be merrily pulling crackers and downing eggnog blissfully unaware of the grave happenings above you and, providing the cold weather stays put, you might not notice anything untoward for quite some time. However, even in the UK all freezing cold weather comes to an end, and with the thaw comes the flood; the ice in your loft reverts back to water and within minutes a deluge of almost biblical proportions is descending from your loft.

There are two ways of stopping this scenario in its tracks. The first is to not insulate your loft, but as the heat goes through the roof so will your fuel bills. The second, cheaper and more sensible approach is to insulate all the pipework up in the loft. A number of insulating options are available:

HESSIAN ROLL

This is a large roll of renewable and natural insulation, which you wind around existing pipework or thread over new pipework as it's installed. The advantage of hessian is that it's a natural product, aside from the plastic sleeve it's adhered to, and that rodents don't seem to enjoy the taste of it that much – I suspect it has a bit too much of the Bran Flakes about it. The downsides are that is takes an age to apply and it's difficult to apply the right amount of insulation evenly around all your pipes.

FOIL-BACKED INSULATION

This is the ultimate in insulation and as such is fairly expensive. Generally it would be unnecessary for home use, but if the pipework is particularly exposed to the cold this might be an idea. It also has the benefit of being fairly resistant to the attentions of rodents and comes in sizes large enough for waste and condensate pipework – see 'Frozen condensate pipework' (Chapter 4). This is applied in much the same way as foam insulation.

FOAM INSULATION

This is the product of choice for most new insulation projects and will be the one we're focusing on here. It has the advantages of low price and ease of use. On the downside it's not the most environmentally friendly of products and rodents absolutely love it!

TIP *'What's with all these references to rodents?' you may be asking yourself. Well, if you live in the city it might not be a problem, but for older homes out in the countryside mice and squirrels can create merry hell within your loft. When they're not electrocuting themselves by eating cables they like to dine out on various items within the loft, and foam insulation is held in high culinary regard by almost all members of the rodent family. As a result you might find that the insulation you applied last year is now nothing more than delicate little piles of plastic foam dotted just far enough away from your pipework as to render them useless as an insulating material.*

When selecting foam insulation you need to know three things:

- What is the diameter of the pipe to be insulated (normally 15mm or 22mm, but occasionally 28mm)?
- How warm does this insulation need to be?
- How much of it do I need?

(If you have to work out the diameter or your pipework, bear in mind that the diameter quoted by the manufacturers and merchants is the *internal* diameter. As such, a 15mm pipe will be a little over 15mm when measured from the outside.)

There's an enormous calculation with lots of numbers, tons of symbols and umpteen squiggles that you can use to determine just how snug and warm the insulation for your pipework needs to be. Fortunately, your local plumbing merchant has taken this formula and condensed it into its most basic form: 'thick' or 'thin'. It's the sort of thing that would make a mathematician cry, but at least we mere mortals now know where we stand.

In an ideal world you'd always opt for 'thick' insulation, for the simple reason that no pipe has ever burst because it was over-insulated. However, thick insulation is almost impossible to apply if two or more pipes are running close together or if the pipes have been clipped to a wall or joist. As it happens, most plumbing pipework is of a sociable nature and it's rare for a solitary pipe to make its way across a loft space, and when it does it will almost always be clipped down to a wall or joist. With this in mind you might be better off opting for thin wall foam insulation.

The foam itself usually comes in 2m lengths, so you need to get a tape measure out and roughly measure the amount of pipework in your loft. Bear in mind that most merchants will take back unused lengths of insulation, so if in doubt buy a bit more.

Applying foam insulation to pipework

Tools and materials
- Mitre block
- Tenon saw
- Tape measure
- Marker pen
- Duct tape
- Knee pads

1 The first stage is to measure the length of pipe you're going to insulate and then cut the insulation accordingly. In this example, we'll say we have a 3m length of straight pipe, which makes the first bit easy as the insulation comes in 2m lengths.

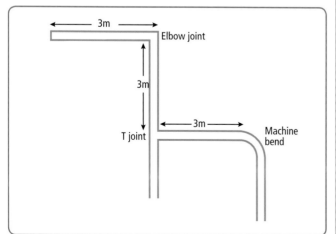

2 To apply the insulation look at the end and find the split that's been put in it, but not put all the way through. Once you've found this you can pull the insulation open, and once you have a small section freed up you can just hold one end and slowly run a finger down the foam to open it up completely.

3 Once you've opened up about a metre of insulation you can put the opened end on to the pipe and start pushing it along the pipework, opening up more of the foam as you go.

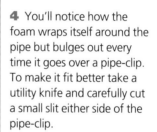

4 You'll notice how the foam wraps itself around the pipe but bulges out every time it goes over a pipe-clip. To make it fit better take a utility knife and carefully cut a small slit either side of the pipe-clip.

5 Now take a piece of duct tape and wrap this around the insulation either side of the pipe-clip to hold everything in place and to make sure no bare copper is left exposed to the elements.

CUTTING CORNERS
We're always told that cutting corners is a bad thing, but not when it comes to insulation. In our example we have 1m from the

end of the insulation to the middle of the 90° bend (also known as an elbow joint), so you measure off 1m on your new length of insulation and put it in the mitre block.

1 Use the 45° angle in the mitre block and make sure your tenon saw cuts across your mark in the middle of the insulation.

2 You now split the insulation as before and slide it on to the pipe until the bend is nestled in the middle of the cut.

3 You now have two flat ends of insulation hard against each other. Use the duct tape to stick these together.

4 The distance from the top of the bend to the branch is 3m, so you can now just cut the other side of your 90° bend by repeating the above steps and push the whole length of insulation on to the pipe, ensuring that the two cut edges fit snugly together.

5 Now tape them together to stop the joints opening up.

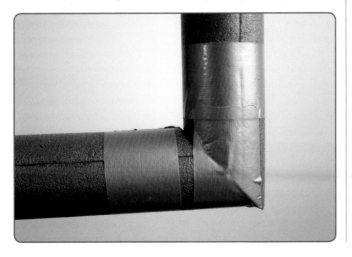

INSULATING A BRANCH

The next stage in our example shows a branch, or tee.

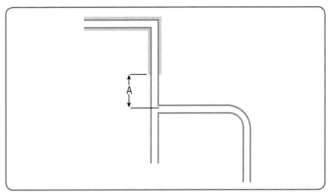

1 To insulate this, measure the distance between the end of the existing insulation and the middle of the branch (distance A in the diagram). Now mark this distance on the tube of insulation and put it into the mitre block.

2 You make the first cut exactly the same as for the 90° bend. Having done that, you then make a second cut straight across the insulation, chopping off the end of the wedge to give you the shape shown.

3 Now fit this piece of insulation and repeat the process for the other side of the branch.

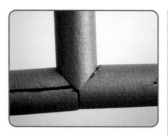

4 To fit the insulation for the branch itself make the standard 45° cut then turn the foam over and make a second reverse cut. The end result should be a perfect V shape, which now fits neatly into the gap to form a perfect foam joint. Use the duct tape to secure everything in place.

INSULATING MACHINE BENDS

The pipe now runs a further 3m before bending down 90° in what's usually referred to as a 'machine bend', as opposed to the more severe elbow joint we just looked at.

Because this next bend is a gentler affair you can get away without cutting the insulation, providing you use a judicious amount of tape to keep it wrapped completely around the pipe.

As you split the insulation and push it over the pipework you'll

notice that it will open up slightly as it goes around the bend. You might also notice that the degree to which it opens will increase or decrease if you now give the foam tube a slight twist. The idea is to reduce any gaps, so twist it accordingly and apply tape to the start, middle and end of the bend to ensure that no bare pipework is left.

And that's it! Just repeat this procedure for all the un-insulated pipework you find, bearing in mind that you must cover ALL of the pipework – gaps in the insulation are just starting points for freezing and subsequent pipe bursts. If you come across valves then insulate them as much as is practicable, bearing in mind that they have to remain usable and vaguely visible if you're ever going to use them.

TIP *As you wander around your loft you may notice things like plastic overflow pipework and possible gas pipework. In theory none of these need insulating but, if in doubt, insulate. No pipe has ever burst because it was over-insulated.*

Insulating your storage tanks

Most people think you insulate your cold-water storage tank to stop it freezing in winter. Whilst this is important the insulation is equally, if not more importantly, stopping your water getting too warm in the summer. The water in an un-insulated tank can reach 30° in the summer, which is an ideal temperature for umpteen bacteria to start growing. Most of these will be harmless but it's still not a good idea to bathe in or drink bacteria-ridden water.

Most homes actually contain two tanks hidden up in the loft. The smaller of these is for the central heating and the larger is for your hot, and sometimes your cold, water. As a child I was always told I shouldn't drink out of the hot-water tap. I never realised why until I had a look inside our storage tank; I've never drunk from the hot-water tap since.

Most storage tanks are insulated and protected by anything that happened to be to hand last time someone was in the loft. So carpets, bin bags full of old clothes and assorted bits of polystyrene often form the bulk of the protection. Whilst there's nothing inherently wrong with this approach it doesn't really protect the tank's contents, and it's really this stored water that you're trying to keep cool and clean. Fortunately most tanks come with a kit designed for just this purpose.

The kit in question is called the 'Byelaw 30 kit' or, if you wish to be bang up to date, the 'Regulation 16 kit'. It's composed of insulation, a close-fitting lid, a rubber seal and two insect-proof filters. Once applied to your tank the water within will be protected from any contaminants as well as from the vagaries of the weather.

TIP *I've mentioned this before but it bears repeating: if your tank is made of galvanised steel, stop hunting for an appropriate kit and start thinking about replacing the entire tank – see 'Fitting a new cold-water storage tank' (Chapter 8).*

Since these kits have 'close-fitting' lids you'll need to buy one that fits your tank, so the first thing you need to find is the label affixed to the tank. If you can't find a label then measure the length, height and width of the tank. Once you have these details nip down to a plumbing store and order your kit.

Fitting a Regulation 16 kit (Byelaw 30 kit)

Tools and materials
- Adjustable spanner
- Drill
- Hole-saw kit
- Utility knife
- Abrasive strip or round file
- Duct tape
- Sponge and/or water vacuum cleaner
- Marker pen

1 First off clear away everything else so that you have a lidless and insulation-free tank. Check that the tank still looks solid and leak-free and check the state of the ball valve and overflow pipework.

2 Part of the job of the Regulation 16 kit is to keep your hot water clean. As such, it's a bit pointless fitting one if the water is already dirty and the tank is filled with sediment. If this is the case drain down the cold-water storage tank by turning off the mains cold water and opening all the hot-water taps in the house. To reduce the risk of an airlock close all the taps just before the water drops below the level of the outlet pipe and use a sponge and bucket to remove the rest of the water – or use a water vacuum cleaner if you have one.

With the water gone give the tank a good clean, but don't use any cleaning products; fresh water, a sponge and tissue paper are fine.

For the smaller F&E tank this is less of an issue, as it's never going to be drunk or bathed in – hopefully. That said, the gunk in this tank could be drawn into your central heating where it could cause a blockage, so it's a good idea to clean it out.

To do this, turn off your CH heating and the cold-water mains. You can empty the tank using either a water vacuum cleaner or a bucket and sponge. Alternatively you could open up a drain cock on one of the radiators (see 'Draining down your central heating' in Chapter 2), but this will draw all the gunk into the system and there's usually not that much water to remove anyway.

With the tanks clean, close all taps and drain cocks, turn your cold water back on and let the tanks fill with clean water.

3 The first thing you need to fit is an insect-proof barrier to the overflow pipe. To do this use an adjustable spanner to loosen the tank connector where the overflow pipe currently connects to the tank.

4 Remove the old connector completely and fit the new insect-proof connector, making sure that the rubber seal is on the outside and that you've tightened it up sufficiently to stop any movement.

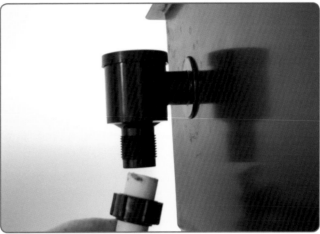

5 Undo the plastic nut on the outside of the new connector, push this and the plastic olive – if one is supplied – on to the overflow pipe and then screw the nut back on to the tank connector, using the adjustable spanner to give it an extra half-turn once you've tightened it up as much as you can by hand.

You might have to adjust the overflow pipework to get it to fit the new connector. If this is the case read 'Solvent weld pipe' in Chapter 6.

6 The overflow connector comes with a dipper. The idea here is that this terminates below the water level and in doing so stops cold winter air blowing down the pipe and freezing the water in the tank. You usually have to attach this dipper – it just pushes on – so do it now and make sure the end is just under the water. If

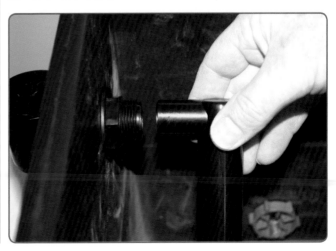

it isn't you need to adjust the ball valve so the water level rises a bit. You can do this by adjusting the height of the ball on the arm using the screw that attaches the ball to the arm of the valve.

Old ball valves don't have these screws so you may have to resort to just holding the arm in both hands and gently bending it up a little.

7 The next thing to fit is the lid, and to do this you need to cut out two holes – one for the vent pipe and another to let air, and nothing else, in and out of the tank. You'll find the vent pipe dipping down and into the storage tank. It's usually a copper pipe and is where air normally escapes from the hot-water system and where water gushes out in an emergency. Measure its distance from the edges of the tank and mark this position on the new lid.

TIP The vent pipe should always drop into the tank and not just hover above it. Sadly this isn't always done properly when the tank is fitted so you might have to extend the vent. The easiest way to do this is to use a 22mm push-fit socket and a short section of 22mm pipe – either plastic or copper. See 'Plastic push-fit pipework' in Chapter 6. If you're going to extend the vent pipe it's important that it doesn't now dip into the water – it needs to just pass through the lid, no more.

8 Your Byelaw kit comes with a rubber bung with a hole in it. This fits on the vent pipe as it passes through the lid and ensures an insect-proof seal. Measure the size of this bung and select an appropriately sized hole-saw. Drill the hole for the vent pipe where you

marked it on the lid. Once you've cut the holes use a bit of abrasive strip or a round file to clean the edges.

Don't cut any of your holes over the tank, as you don't want debris getting into it. The plastic shards have a habit of being drawn into the pipework and blocking it, usually years later just at the base of a tap.

9 To fit the bung it's usually easier to slide it on to the vent pipe first, and then push it down into position when the lid has been fitted. A dab of silicon grease makes it easier to push the bung into place.

10 You now need to fit the air vent. This can go anywhere in the lid so just pick a vaguely sensible spot. Measure the base of the air vent and drill an appropriately sized hole in the lid.

11 Push the air vent into position. If it seems a bit loose you can use a bit of solvent glue to hold it in place.

12 Fit the lid in place, making sure the vent pipe passes through the appropriate hole. If the vent pipe is too long it can be close to impossible to fit the lid. If this is the case cut the pipe back a bit. The vent pipe should under no circumstances touch the water. It must terminate below the top of the tank but at least 44mm (the internal diameter of the pipe x 2) above the top of the overflow outlet.

13 Push down all the lid edges to ensure it's firmly in place.

14 We're now ready to wrap the insulation around the tank. To help with this the kit normally comes with a length of plastic

thread. Personally I find this little or no help, so I use a bit of duct tape to stick one end of the insulation to the tank, then pass the insulation around the tank and use another bit of duct tape to secure it in place.

You'll have to cut holes and slices in the insulation where pipework enters and leaves the tank. Just use a bit of duct tape around the holes to keep everything in place. If you like you can now wrap the plastic thread around the tank just to be sure.

15 The only item now remaining is the insulation for the lid. You

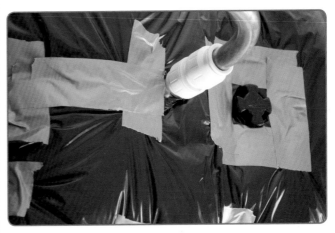

need to cut out a hole in this for the air vent, and you'll need to slice it open so that you can wrap it around the vent pipe, securing it back in place with a piece of duct tape.

16 Just to make sure everything stays in place you may now like to attach the insulation on the lid to the insulation around the tank with some more duct tape. Don't go over the top with this as you may need to gain access to the ball valve in the future.

So there we have it – a supply of perfectly clean water that won't freeze in the winter nor get unhealthily warm in the summer.

Home Plumbing Manual

6 PIPEWORK AND BASIC TECHNIQUES

Plastic push-fit pipework	53
Tectite metal push-fit pipework	55
Copper compression pipework	56
Using jointing compounds and tape	58
Removing compression fittings	58
Using compression joints with plastic pipework	59
Using chrome pipe	60
Soldered pipework	60
Black iron or steel pipework	63
Lead pipework	64
Waste pipework	65
Name that fitting	68
Bending pipework	70
Pipe sizing	72
Supporting pipework	73

The word 'plumber' derives from the Latin *plumbum*, meaning lead, and refers to the fact that for millennia most water was run through lead pipework. In fact everyone from the ancient Egyptians through to the Victorians used lead for all sorts of things, from water containers to make-up. Sadly this is no longer true. I say sadly because lead is a lovely metal to work with, but has one huge drawback – it's very toxic, which possibly explains why you don't see many ancient Egyptians around any more.

If you find any lead water pipework in your home you need to get it replaced. If this isn't possible then fit a water filter that removes lead to at least one tap and take all your drinking water from this. However, these days most of the pipework you'll find in your home will be copper, steel or plastic, and if you're going to do any real plumbing you're going to have to deal with what's already there, as well as learn how to add your own bits. So let's have a look at the options:

Plastic push-fit pipework

This is the easiest type of pipework to work with and as such is the one I'd recommend for beginners. There are a number of different types out there, the main ones being Speed-fit, Hep2O and In4Sure – although they seem to be moving towards calling this 'the latest Hep2O'. A number of DIY stores also sell variations on a theme so don't be surprised if you discover plastic pipework that's slightly different from the ones we'll be discussing.

SPEED-FIT

This is produced by John Guest and is probably the most common plastic push-fit system you'll come across.

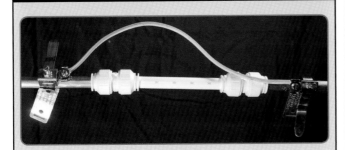

⚠️ **WARNING**

The disadvantage with all plastic pipework is that if you add it to a system largely composed of copper tube you'll have to use earth bonds to maintain the electrical 'continuity'. What this means is as follows: if you drop a live electrical cable on to a radiator then the radiator and every bit of metal pipe connected to it is now electrically charged. If someone then touches a radiator on the other side of the house they'll get a shock, possibly a lethal shock. To prevent this, all the metal pipework in your home will be earthed ... unless you've just added 3ft of non-conductive plastic pipework into the system. So to ensure that you're not breaking the earth, always add earth bonds on either side of the plastic section. Alternatively, ask an electrician to test the system after you've completed your work to ensure that the metal pipework is still earthed correctly.

Another issue with plastic pipework and fittings is that rodents love it! For some reason they seem to prefer the fittings to the actual plastic pipe, but at the end of the day they'll eat it all given half a chance. You need to bear this in mind if you're thinking of using plastic pipe systems in areas where rodents might decide to take up residence.

USING SPEED-FIT

Speed-fit pipe is white and comes with little blue chevrons on it every inch or so. The idea behind this is that if you cut the pipework on one of these marks the next mark will be right on the edge of the fitting when it's been fully pushed in.

Of course, it's not always possible to make your cut right on one of these marks so you should always check to see how far away from your cut edge the nearest mark is and then check that the distance from the end of the fitting to the next mark is roughly the same once you've pushed it on fully.

1 To work properly plastic pipe needs a nice, straight, clean cut and to achieve this you need to buy a pair of plastic pipe cutters. As we saw in Chapter 3, there's a range of different designs but they're at heart a very sharp pair of scissors – although don't try scissors, as they don't work!

2 Once you've made your cut you'll need to strengthen the end of the pipe so that it grips the fitting properly and doesn't buckle under pressure. To achieve this, the system comes with what John Guest refer to as a 'super-seal pipe insert'. You just push one of these into the end of the pipe.

3 You're now ready to push your fitting into place. First, make sure the fitting is in the 'unlocked' position where you can see a gap between the screwcap and the body of the fitting, as opposed to 'locked', where the screw cap is tight to the fitting and there is no gap at all. If it isn't unlocked then just turn the screwcap anticlockwise until you feel some resistance and can see this small gap.

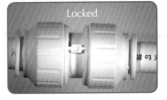

4 With the fitting 'unlocked', push it gently over the pipework. You'll usually feel a little resistance as the pipe reaches the O-ring in the fitting. At this point you push a little harder and you'll feel the fitting slip into place. You can check it's in correctly by looking to see where the next mark is on the pipework. You might find it easier to push the fitting on if you dampen the pipe and fitting with a little water first.

5 Once you've connected up the pipework you 'lock' the fitting by twisting the screwcaps clockwise until there's no longer a gap between the screwcap and the body of the fitting. Now just give everything a tug to check that it's all connected properly.

To be extra safe, *eg* if you have children in the house, you can also fit a clip or 'collet' over the end of the fitting which has to be removed before you can pull the pipe out.

DISCONNECTING A SPEED-FIT FITTING

Sadly, when it does this the plastic super-seal insert is usually left behind in the fitting, and to get it out you need to dismantle the fitting entirely.

1 To disconnect you'll first need to 'unlock' the fitting by turning the screwcap anticlockwise until you can see a small gap and feel some resistance. If clips have been added you'll have to remove these as well, using a small flat-headed screwdriver to flick them out.

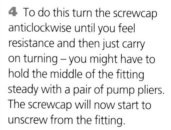

4 To do this turn the screwcap anticlockwise until you feel resistance and then just carry on turning – you might have to hold the middle of the fitting steady with a pair of pump pliers. The screwcap will now start to unscrew from the fitting.

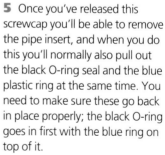

2 Now press down on the collet at the end of the fitting. You can do this with your fingers, but you can also get a little plastic 'collet release' tool if you need one.

5 Once you've released this screwcap you'll be able to remove the pipe insert, and when you do this you'll normally also pull out the black O-ring seal and the blue plastic ring at the same time. You need to make sure these go back in place properly; the black O-ring goes in first with the blue ring on top of it.

3 Now just pull and the pipe should come away from the fitting.

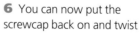

6 You can now put the screwcap back on and twist it clockwise. Once again you'll meet a bit of resistance – give it another twist to get over this and the screwcap is on again.

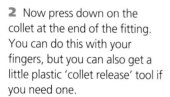

The great thing about these fittings is that they work just as well with copper pipework, the only difference being that you don't get any little markers to tell you if the fitting is in properly and you don't need the pipe inserts to support the copper tube.

John Guest do hundreds of different fittings for this system and most plumbing merchants will have a free booklet you can take away listing all of them. Alternatively look up 'John Guest' on the Internet; their website lists all their products.

Hep2O

Hep2O was used by an awful lot of building companies and as such you might find quite a bit of it in your home already. The easiest way to identify it is by its colour – a rather grim shade of grey. It works pretty much the same as the Speed-fit we discussed above, in that you have markings on the pipe to indicate the depth the fitting needs to be pushed in. You push a pipe insert into the end of the pipework – a metal insert in this case – then you just push the pipe into the fitting.

The big disadvantage of Hep2O is that you can't dismantle the fitting once it's on the pipe; once it's on, it's on! If you absolutely have to dismantle it then you can unscrew the end of the fitting and pull the pipes apart but the metal grip ring and the head of the fitting will remain on the pipe, only to be freed if you're really determined and have a go at it with a hacksaw. In short it's a right pain, and this is why it's no longer the world leader in push-fit systems.

In4sure

In4Sure is produced by Wavin and is effectively their updated and improved Hep2O. In fact, after initially launching it as In4sure they now seem to be calling it 'Next generation Hep2O'. Whatever they're calling it, it does seem to have all the advantages of the Speed-fit system and none of the disadvantages of the old Hep2O. In addition it's tried to address a few other issues.

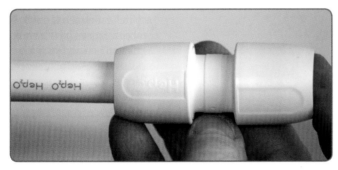

Firstly they've got around the issue of the pipe inserts pulling out and getting stuck in the fitting when you dismantle the system by putting little barbs on the inserts so that they stay in the pipe.

Sadly these inserts are nigh on impossible to remove from the pipe, which can be a problem if you just want to trim the pipe back a bit, but on the whole it's a real improvement.

They've got around the problem of not knowing if you've pushed the pipework in far enough by manufacturing these pipe inserts with a knobbly end. The idea is that you push the pipe into the fitting and then give it a twist. The knobbly bits will then rub against the end of the fitting with a distinctive 'rumble' that you can feel; feel

the rumble and it's in properly, don't feel it and you need to push it in further.

To allow you to dismantle the pipework there's a little ring at the end of the fitting that you press down to release. To stop anyone accidentally releasing it the ring is recessed a little so

that you really need a little tool to depress it sufficiently to release the fitting. They of course supply this tool and you of course will lose it; such is life.

TIP *Although it's generally not encouraged, all the plastic pipe systems are pretty much interchangeable with each other, although you should always aim to use the pipe insert that goes with the fitting. As such you can take copper, fit Hep2O to it then put a Speed-fit fitting at the other end, change to Speed-fit pipe and finally end with an in4sure fitting. I can't imagine why you'd want to, but it's nice to know that you could if the feeling ever took you.*

Tectite metal push-fit pipework

To circumvent the problems with plastic pipework – earthing and providing a meal for the local rodent population – you might want to consider the Tectite push-fit system. This works in much the same way as the plastic systems we've already discussed but because it's metal it maintains earth continuity and leads to large dentistry bills

if attacked by squirrels and their ilk. Tectite do two types of fittings – those that can be released from the pipe using a little tool (demountable) and those that are permanent fixtures (pains in the posterior). Personally I'd opt for the former, unless there's a really good reason not to.

The only downside to Tectite is the cost.

TIP *To convert from plastic to Tectite just put copper tube into your plastic fitting and away you go.*

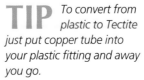

Copper compression pipework

You might as well learn to like compression fittings because if you're going to do anything beyond very, very basic plumbing you'll have to deal with them. The type of fitting you'll generally come across is called 'Type A'. The other type, cunningly referred to as 'Type B', is more involved and is only really used for underground pipework – 'Type A' should *never* be used for underground pipework.

A 'type A' compression joint works by squeezing a little ring of brass or copper (known in the UK as an 'olive') between the fitting and the pipe. Because the pipe is relatively soft copper and the olive is also made of a soft metal everything gets squeezed together until it forms a perfect metal-to-metal watertight seal.

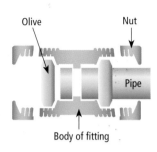

Olive   Nut

Pipe

Body of fitting

TIP *In the bad old days only one edge of an olive was bevelled, but these days both sides are identical. I only mention this because a lot of plumbing books still go on about ensuring the olive is the 'right way around', which tends to confuse people holding a perfectly symmetrical olive in their hands.*

USING A COMPRESSION FITTING

1 Whilst it's not essential, the easiest way to use a compression fitting is to undo the nut at the end of it and slip this on to the pipe, then slide the olive on and finally push the end of the pipe into the fitting.

2 Now push the nut back down on to the fitting and screw it to the thread by turning it clockwise.

3 Once you have it finger-tight grab two adjustable spanners. Put one on the fitting itself and the other on the nut. Now turn the nut whilst keeping the fitting itself steady and in place. Give the nut between three-quarters and one full turn.

TIP *The most useful adage in plumbing refers to compression joints and goes like this: 'Righty tighty, lefty loosey'. It might not be Shakespeare but it's darned useful and I still whisper it to myself from time to time.*

In some instances you'll hear the olive squeaking as it's crushed into position but sadly this isn't always the case. If you do hear a squeak, give it an extra quarter-turn and you should have a perfect watertight seal. If you don't hear a squeak, tighten it until it's going to take some effort to give it another quarter-turn. What you don't want to do is over-tighten it. If you do you end up squeezing the olive until it starts to crush down on the pipe,

at which point everything stops being watertight and starts leaking again, with the downside that you can't stop the leak by tightening the joint a little more.

CREATING A TEST-RIG FOR YOUR FITTINGS

If you're going to be even vaguely adventurous with plumbing you need to get a feel for just how tight you need to turn a compression joint. To learn this you need to practise and test, and an easy way to do this is as follows:

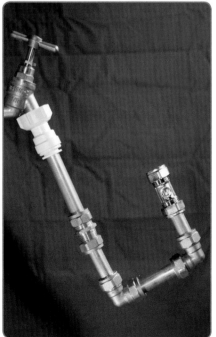

1 Find an outside tap. These are called 'hose union bib taps', and it's the hose union bit that we're interested in as it lets us screw far more than just hoses to the tap.

2 Take the existing hose connector off to reveal the thread at the end of the tap.

3 Buy a ¾in to 15mm push-fit tap connector and screw it on the end of the outside tap.

There are a number of different types of push-fit tap connectors and sadly only some screw on to outside taps properly. Your best bet is to ask for a range of them and see which one fits best – the all-plastic one shown works a treat.

4 Make up some joints however you see fit. In the photo I've gone all exotic and used a straight coupler and two elbows. You can start off with a single joint or go wild and create a pipework masterpiece using whatever fittings are to hand. To make life easy try to end up with just two open ends to your creation.

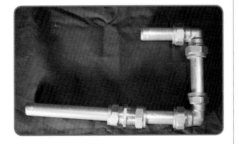

5 Assuming you have only two open ends you need to use one to connect to your tap and want to close off the other using either a 'push-fit stop-end' or an isolation valve. The isolation valve is best and can be bought in push-fit or compression form depending on your mood.

Put your valve of choice on one open end of the pipework and push the remaining open end on to your tap.

6 If you've opted for an isolation valve you need to make sure this is shut by turning the groove in the middle so that it's across the flow, *ie* at right angles to the pipe.

Use a flat-headed screwdriver to rotate the groove until it's at right angles to the pipework, which means the valve is now closed.

If you used a stop-end you don't need to bother with all this.

Open   Closing   Closed

7 Now turn the tap on! If you've tightened everything sufficiently all will be quiet and dry. If you haven't, the ominous sound of dripping water will assail your ears ... or will it?

If you have a very small leak nothing will appear at first because there's air trapped in the pipework, and until this has escaped through the leak no dripping water will appear. It's for this reason that the isolation valve is preferable as you can open it a little, let all the air be pushed out, close the valve, dry everything with tissue paper and see if it stays dry.

If you used the stop-end you can try listening to the hiss of escaping air or just leave the pipework 'on test' for an hour or so to let any air escape and the telltale drips of water to appear.

8 To check for leaks, run a finger around each joint starting at the top and working your way down. Cold pipework will often feel wet even if it isn't leaking, so having wiped your finger around the joint, have a look at it – is it actually wet? If it is, that's where the leak is; if it's still dry continue down the pipework until you find the leak. To stop the leak just give the nut on the fitting a little tweak, *ie* turn it clockwise a fraction – and most of the time it really does only need to be a tiny fraction, say a tenth of a turn, to stop the leak dead in its tracks.

Repeat this entire exercise a few times until you feel you have an idea of the amount of pressure you need to apply to a compression fitting, always remembering that whilst you can always tighten a joint that's been under-tightened, you can do little about a joint that's been over-tightened.

You can use this testing set-up to test any type of fitting, so play around until you're confident that you can create dry pipework first time every time.

Using jointing compounds and tape

There's a school of thought that says you should always put either PTFE or 'jointing compound' around the olive before tightening the fitting. Personally I don't subscribe to this idea for the simple reason that on new or undamaged copper a fresh olive works perfectly well by itself, and why fix something that isn't broken? However, there are a number of scenarios where a bit of help might be appropriate. These are:

- Reusing an old olive with a new fitting.
- Putting a new olive on to pipework that's been compressed slightly by an old fitting,
- Fitting a compression joint in an area where it'll be a nightmare to get at it again once everything is finished.
- Where the joint is leaking, despite (or possibly because of) your best efforts to tighten it.

In these circumstances there are two approaches you might take.

The first approach is to take a length of PTFE tape and wrap it around the olive about three to four times, then retighten the joint. The PTFE will help the olive form a waterproof seal even if it's previously been over-tightened.

Alternatively you could use a jointing compound rather than tape. The advantage of this is that the compound is a paste and as such is often easier to apply than tape. To apply the jointing compound just put a small amount on your fingertip or a small brush and smear it over the olive, making sure you get it all the way around the joint.

Be aware that there are a number of jointing compounds that are fine for central-heating systems but can harbour bacteria that can infect drinking water. So when using jointing compounds, be sure to check

that they are safe to use for drinking water – or 'potable water', as it's often called. Jet Blue is a popular jointing compound for potable water, whereas Jet White can only be used for non-drinking water.

Plumbers' hemp also used to be used extensively as a washer or as a jointing compound. However, this too encourages bacteria to grow in the fitting, so shouldn't be used on potable water. To be honest the smell of plumbers' hemp alone should be enough to ward you away from using it on drinking water. I've no idea why, but it has more than a hint of old horse about it.

TIP *Remember that with compression fittings it's the meeting between the pipe, the olive and the fitting that creates the waterproof seal. You may come across leaking compression joints encased in half a dozen rolls of PTFE tape, all wrapped around the thread of the nut, and all utterly useless because that's not what's stopping the water getting out – always apply the tape to the olive.*

Removing compression fittings

The problem with compression fittings is that, whilst it's easy to undo the actual fitting, it still leaves you with a nut and olive firmly attached to the pipe; a nut and olive you'd much rather remove altogether. There are, however, a number of ways of removing the olive:

OLIVE REMOVAL TOOLS

There are two basic types: one that cuts into the olive to allow you to remove it, and another that pulls the olive off. The advantage of these tools is that they remove olives quickly and reliably with little danger of damaging the pipe. The olive pullers will usually work with a number of different pipe sizes but struggle to remove olives that have been over tightened. The disadvantage of both tools is that they're expensive for what they are, so much so that I've always been far too tight to buy one.

PULLING THE OLIVE OFF

There is a cheaper approach that works the same way as the olive puller:

1 Undo the fitting you wish to remove.

2 Put a set of pump pliers on the nut.

3 Draw the pliers up to the olive and then pull and waggle it until the olive and nut come off. Sometimes this works, sometimes it doesn't, and sometimes you end up damaging the pipe.

CUTTING THE OLIVE OFF

 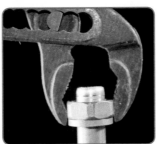

1 An alternative is to take a junior hacksaw and carefully cut into the olive. It's important not to cut into the pipe itself, so aim to go *most* of the way through the olive without going *all* of the way through.

2 Now insert a small flat-headed screwdriver into the groove you've cut.

3 Twist the screwdriver to pop open the olive. This will only work if you've cut deep enough into the olive, so if nothing happens saw a bit deeper and try again.

4 You can now just pull the olive off, being careful not to cut yourself on the sharp edges.

The advantage of this approach is that it involves no expensive tools; the disadvantage is that it can take a fair length of time to get the olive off, and if you cut into the pipe you're stuffed.

Using compression joints with plastic pipework

A frequent scenario is opting to use a plastic push-fit system but then finding that you need to connect to a compression fitting. Of course, you could just replace the compression fitting with a push-fit equivalent but this isn't always possible. So how do you use plastic pipework with a compression joint?

1 Firstly, dismantle the compression joint and push the nut on to the plastic pipe.

2 Now push the olive on to the pipe.

3 Put a pipe insert into the end of the pipe and push it fully in.

4 Push the end of the pipe into the fitting – you might have to give it a real push, as, depending on what system you're using, the pipe insert may have a little rubber seal on it which will create some resistance but will also help keep everything dry.

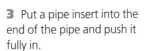

5 Now push the nut and olive down and tighten the joint as usual.

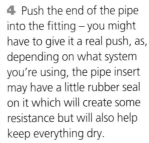

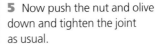

TIP *The most important part of this sequence is the pipe insert, as without it the plastic pipe will just crush as you tighten the olive and you'll discover a world of water when you turn the mains back on. Also, the order in the above sequence is critical as it's nigh on impossible to get the olive on to the pipe once the pipe insert's in.*

Using chrome pipe

I hate chrome. I really do. It might look nice, but it's a pain in the derriere to work with. It's usually just plain old copper tube with a thin plating of chrome on the outside, but the problems all have to do with the chrome itself; whilst it might look lovely it's very, very hard.

As a result of this hardness, compression fittings often don't work because the olive can't compress the pipework, so it's can't grip it properly. And push-fit fittings no longer work because the steel grippers inside the fittings can't cut into the chrome, and so slip off. And finally, solder doesn't work because the chrome can just peel off the copper at a later date and cause a leak.

Of course, when I say 'doesn't work' what I mean is 'doesn't work in the long term', because all fittings will give the appearance of working perfectly well with chrome ... right up to the moment the fitting pops off and the deluge begins.

So if you insist on using chrome pipework, you have to make sure you remove the chrome from the end of the pipe before you push it into a fitting.

1 Push the chrome pipe into the fitting to be used and mark where it emerges by wrapping a length of masking tape around the end of the fitting.

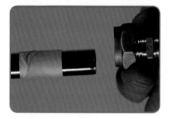

2 Now remove the fitting and apply a little more masking tape just a few millimetres closer to the end of the pipe so that you'll only see chrome when you've finished.

3 Now set aside a fairly considerable period of time whilst you gently run a flat file around the chrome, slowly but surely removing it and exposing the nice, clean copper underneath.

4 Once you have clean copper all the way around the pipe give it a quick rub down with an abrasive strip to make sure it's nice and smooth. Now remove the tape and your chrome pipe is finally ready to be used in a fitting.

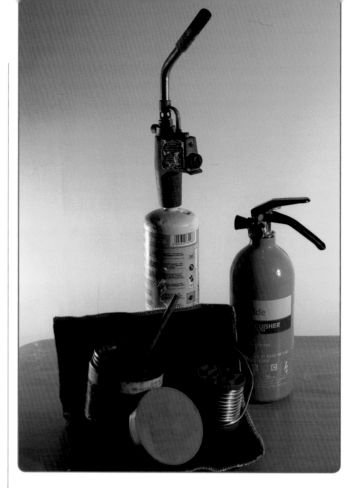

Soldered pipework

If you're feeling really adventurous you might want to use soldered copper joints with copper pipework. The advantage of this approach is that the joints tend to be more robust and last longer than any of the other methods, and they're also far cheaper – a real plus if you're intending to fit a lot of pipework.

Before you start, though, you'll need to acquire some additional equipment:

- Blowtorch or pipe soldering iron
- Heat mat
- Solder
- Flux and flux brush
- Fire extinguisher
- Abrasive strips

BLOWTORCHES

There are two ways to heat everything up: a blowtorch, or an electrical pipe-soldering tool. The former is difficult to use well, and dangerous to use at the best of times. The electrical soldering tool is difficult to buy but easy to use. On the downside it takes far longer to heat the fitting up to the correct temperature and you have to drag around an electrical lead everywhere you go. Most professional plumbers will use a blowtorch, but you can very easily burn your entire house down if you make the slightest mistake, so I wouldn't recommend it unless you've set aside the time to practise a fair amount beforehand.

Most blowtorches allow you to adjust the flame from high to low. When set to low the end of the blowtorch nozzle can often become horrendously hot, so be very careful where you lay the torch down afterwards.

HEAT MAT

Sod's Law insists that the pipework you want to solder will be lying very close to something that's just dying to burst into flames; wallpaper is a prime example. To prevent fire and burn marks you need to purchase a heat mat – this is far less important if you're using an electric pipe-soldering tool. They come in a variety of types and colours, but all are made of fireproof material. You slide the mat between the pipe you're heating and the surface that the blowtorch might damage.

Most, if not all, heat mats won't cope with heat fired directly at them. Whilst some might say that a heat mat that can't cope with heat would be more aptly described as a mat, the fact

remains that you need to try to fire across the mat rather than straight at it.

You can buy heat mats that have a hole built into them so that you can wrap them around pipework. You can also buy a spray that absorbs heat, which can be extremely useful to have to hand.

TIP *Copper is very conductive of heat, so whilst you'd normally aim to heat only the fitting, it's possible to move the flame farther away and let the copper pipe conduct the heat into the fitting. In this way you reduce the chances of damaging combustible materials around it.*

SOLDER

Solder is an alloy of metals designed to melt at a relatively low temperature – at least, low by comparison with most other metals (usually just under 200°C). Because of the disparity between the melting points of copper and solder you can heat up a copper fitting and its pipework so that it's hot enough to melt the solder without harming the pipework itself. Once you've reached this temperature you apply the solder, which promptly melts and runs into the fitting.

In the old days all solder was made from a lead alloy. They then discovered that lead was poisonous and figured that we probably shouldn't be drinking water that bubbled out of lead pipes or emerged from copper pipes stuck together with lead solder. So lead solder was out and lead-free solder (usually made from a silver

alloy) was in. Nevertheless, you can still buy lead solder, as it can still be used on CH pipework.

However, life is easier if you *always* buy lead-free solder. Not only can it be used on anything but it also acts slightly differently to leaded solder, so you might as well get used to the safe stuff.

FLUX

It's desperately important to ensure that solder runs all the way around the pipe and deep into the fitting, and to make sure this happens you need flux. Flux is a paste that you apply to the fitting and the pipework before you start heating it up. Its primary role is to ensure that the solder runs freely into the fitting, but you can also buy superfluxes, which contain an acidic cleaning chemical that removes any oxidised metal from the surface of the copper.

You can always tell if you've forgotten to flux a pipe because the solder tends to just melt and form a blob. Another sign of forgotten flux is that everything leaks like a sieve.

FIRE EXTINGUISHER

Over the years there have been many fine ideas: the aeroplane, the TV, computers, lemon sorbet. Sadly, using a blowtorch in a flammable home with all the water turned off doesn't make the list.

If you're going to use a blowtorch or an electrical heat source to solder pipework you've also got to have something that can deal with any resulting fires, and the best option is a small fire extinguisher. It doesn't really matter what type you opt for as you're unlikely to be soldering chip fat or electrical cabling, but the foam ones probably make the most sense.

USING INTEGRAL SOLDER RING FITTINGS

There are two main approaches to soldered pipework: the easy way and the not so easy way. The easy way is known as 'integral solder ring' fittings, or often just 'Yorkshire fittings' after the name of the company that invented them. An integral solder ring fitting is one where, instead of applying solder to the end of a fitting and waiting for it to run into it, the manufacturer has produced a fitting with a ring of solder already in it. As such, all you have to do is heat up the fitting and wait until the solder melts.

1 Take a piece of abrasive strip and clean up all the pipework so it's nice and shiny.

2 Use a round file to remove any burrs from the cut end of the pipe.

3 Take a flux brush, or any small brush, and dip it into your pot of flux. Now run the flux around the inside of the fitting.

4 Flux the last centimetre or so of the pipe.

TIP *There's a theory which states that flux should only be applied to the pipework, and not to the fitting. Fluxes can corrode pipework if they're not washed away afterwards; and the idea is that by applying it only to the pipework, you ensure that any excess flux is pushed out of the joint rather than into it. On the downside you're also increasing the likelihood of the solder not running into the joint properly, so I'd recommend using a brush and applying a small amount to the pipework and the fitting. Once the pipework is in operation the hot and cold water flowing through it should remove any excess flux.*

5 Fit the fitting on to the pipework. Never solder just one end of a fitting – always make sure it's connected to all its pipework and solder the entire fitting at the same time.

6 Heat up the pipework. We're using a blowtorch here but you might want to opt for the safer electrical pipe-soldering tool.

The temperature is usually about right when the copper starts to discolour. Shortly after this you should see a tiny line of solder appear around the end of the fitting.

7 Stop heating the pipework and check that this line of solder goes right around the fitting.

8 Leave the joint to cool down then clean it with a damp cloth to remove any flux residue. All you should be left with is shiny copper with a tiny ring of solder visible around the fitting.

TIP *Old copper is darker and duller than new copper because its surface has started to oxidise or rust. Soldering to rust isn't a good idea, so before you solder copper you need to ensure the pipe is clean and shiny. Whilst an abrasive strip will do this for you, you can achieve the same result with a superflux. These contain a mild acid that removes all the oxidised copper and ensures a good soldered joint. On the downside you shouldn't use them for pipework that isn't going to have flowing water in it because the acid also corrodes the pipe if it's not flushed away. It also stings like merry hell if you get it in a cut.*

USING END-FEED FITTINGS

The more difficult approach to soldering is to use 'end feed' fittings. The reason that most professional plumbers use them is because they're also the cheapest way to fit pipework together and, when done properly, give a more robust and longer-lived joint. To use it you start off by following the same steps as for an integral soldered ring fitting:

1 Clean the end of the pipework using an abrasive strip.

2 Apply flux to the inside of the fitting and to the last centimetre or so of the pipe using a flux brush.

3 Fit the fitting on to the pipework – you want to solder the entire fitting at the same time, so connect up all the pipework going into it.

4 Here's where it gets different. Prepare your solder by unwrapping about 10cm of it, straightening it out and putting a little bend in it about 2cm from the end.

5 Use a heat gun or blowtorch to heat up the fitting until you see the copper start to change colour.

Holding the solder roll, apply the tip of your length of solder to the very edge of the fitting. If you're using a blowtorch try to make sure you do this on the side of the fitting farthest from the torch, the idea being that if that side's hot enough to melt the solder,

then the whole of the pipe is hot enough and the solder will melt all around the joint.

If the temperature is high enough the solder will begin to run into the fitting. You want to apply an amount roughly the same as the diameter of the pipe, and this is where that little bend helps – we set it about 2cm from the end, so when the bend has almost vanished we've added enough solder to a 15mm fitting.

6 Take the heat away and let everything cool, then clean away the excess flux using a damp cloth.

 **TIP** *One of the hardest parts of soldering to get right is recognising when the pipe is hot enough. To judge this you need to look out for a colour change:*

A cool pipe.

The pipe is starting to change colour – apply the solder and see if it starts to melt.

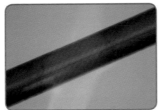

Definite colour change. Remove the heat and the solder should melt as soon as you touch it to the fitting.

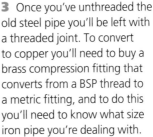

Very strong colour change. You've overheated the pipe, the flux will probably have burnt off and the solder won't flow properly. Best to let it cool, apply more flux and start again.

Regardless of how you choose to join all your pipework together you must remember one very important fact – all approaches, whether push-fit, compression or solder, rely on the fact that the pipework going into the fitting is absolutely, perfectly, round. So before you put your pipework into the fitting check that there are no dents or blemishes. When using new pipework I tend to cut off and throw away the last centimetre of tube as this is where the pipe's most likely to be deformed.

Black iron or steel pipework

In the 1970s copper became so expensive and so scarce that many people gave up on it altogether and started running their water through iron pipework. As such many homes from that era will still have iron pipework in them, so we might as well discuss what to do about it.

First off, working with iron and steel pipework is awkward at the best of times, so you're best off converting it to copper or plastic pipe as soon as you can. The easiest way to do this is to find where the pipework is jointed – they're usually fitted together using threaded joints – and undo the pipe at this point.

1 To unthread the joint, you're best off using a Stilson wrench. To use this type of wrench open the jaws using the adjustment ring until they can be loosely pushed over the pipe.

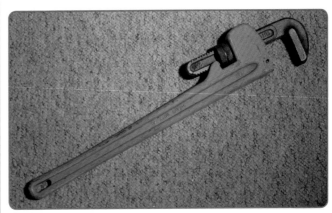

2 You'll need to turn them anticlockwise to open the joint, and if you have the wrench the right way around the jaws will clamp down tighter and tighter the harder you pull. If you have the wrench the wrong way around the jaws will just slide around the pipe uselessly. If this happens turn the wrench over and try again. If the joint is refusing to open, either find a few friends to give you a hand pulling the wrench, heat up the joint using a heat gun or blowtorch, or freeze shock the joint using a proprietary spray.

3 Once you've unthreaded the old steel pipe you'll be left with a threaded joint. To convert to copper you'll need to buy a brass compression fitting that converts from a BSP thread to a metric fitting, and to do this you'll need to know what size iron pipe you're dealing with.

In many cases the fitting will actually have the size printed on it, but if not you'll need to measure the diameter of the inside of the joint, which in a domestic home is usually either ½in, ¾in or 1in, which in new money converts to 15, 22 and 28mm respectively. Once you

have the thread size the only other issue is do you need a male or a female connector? I'm assuming that most of us know what this means but for those who've led a very sheltered life, here's a photo.

1 in pipe

Brass female connector

Brass male connector

Now you can pop down to your local plumbing merchants and ask for the right connector, ie a 15mm to ½in BSP male connector or a 22mm to ¾in female connector.

4 In this scenario there are no olives and no washers – it's the thread that makes the joint waterproof, and to ensure that happens you need to wrap the thread in PTFE tape, remembering to wrap it clockwise around the thread. Most manufactures say four wraps are enough, but most plumbers wrap it a bit more

just to be sure, and some only stop when they run out of tape. Don't be daft about it but make sure the thread is completely covered.

5 To make absolutely sure I tend to also apply a bit of jointing compound – ensuring first that it's suitable for drinking water. Just smear a little dollop around the thread.

6 I also smear a bit inside the fitting, just to be safe.

7 Now screw the connector into the iron pipe joint using an adjustable spanner to tighten it.

8 The other end of the connector is now a standard compression joint. So fit the appropriately sized length of copper into this, tighten it up as described earlier and away you go!

Lead pipework

The first thing that should cross your mind when you find lead pipework is 'Is it feasible to remove this pipework entirely and replace it with plastic or copper?' You really ought to try for an answer approximating to 'Yes,' because it really isn't a good idea to drink leaded water. Having said that, it may be far too expensive to replace all the lead, at least in the short term. If this is the case you'll have to make do with reducing the amount of lead and making adjustments in lifestyle to accommodate it in relative safety.

To reduce the lead you need to trace the lead pipe back as far as you can and then convert to either plastic or copper.

1 Having found this location take a hacksaw and, having made sure that the pipe is empty of water, cut through it, trying your best to get a clean, square cut.

2 Having made the cut you need to clean up the pipe to accept a fitting. Start off with a flat file to get rid of the worst of the blemishes and dirt.

3 Now use an abrasive strip to get the pipe round and really clean.

4 Use a round file to clear the inside of the pipe of any burrs and deformities. Don't overdo this, as the film of limescale coating the inside of the pipe is stopping any lead getting into the water.

5 You now need a fitting known as a Lead-Lok, but first you need to know what size lead pipework you have. Once again it's the internal diameter you're looking for, and with lead it will always be in imperial measurements, normally ½in, ¾in or 1in. (When converting, ½in lead normally corresponds to 15mm copper, ¾in will be 22mm and 1in either 22 or 28mm.)

To add a bit more confusion lead was also sold by weight, so you could have ½in 6lb or 7lb lead. So measure the cut pipe, pop down to your local plumbing merchants and buy every lead fitting they have that looks about right – just make sure they're OK with you bringing back any you don't use. An even easier approach is to take a bit of the lead you chopped off earlier and ask the merchants for a Lead-Lok that fits it.

Calling such fittings a Lead-Lok is like calling a vacuum cleaner a 'Hoover', *ie* it's the name a single manufacturer gives to its specific product. However, most merchants will know what you mean if you ask for a Lead-Lok, even if what they hand over is called something different, *eg* a Leadline.

6 According to the manufacturers you can just push the fitting directly over the pipe, without the need to dismantle it, but

personally I play safe and dismantle the fitting first. So first put the nut on to the pipe.

7 Now push the split olive on to the pipe, ensuring the that flat face points towards the fitting.

8 Then push the rubber O-ring on to the pipe.

9 Finally push the fitting on to the pipe, ensuring that the pipe is fully inserted into the fitting.

10 Push the nut back down the pipe and tighten it by hand before using a pair of adjustable spanners or plump pliers to give the nut an extra three-quarter or full turn. Remember, this is lead you're tightening on to, a metal not renowned for its brittle strength, so there's no need to overdo the tightening.

11 You now have a standard copper compression fitting on the other end and can continue the rest of the pipework in either copper or plastic.

⚠ **WARNING**

If you have lead pipes and can't get rid of them altogether you should always try to run the taps for a while before drinking out of them, or, better still, buy a water filter that removes lead and fit it to a single tap that can then be used to supply your home's drinking water.

Waste pipework

Most waste pipework is now run in plastic, although you will occasionally come across cast iron, lead and copper.

Sadly, most manufacturers of waste pipework don't seem to get on; if company A sells a pipe called '32mm waste' you can bet your bottom dollar that it won't fit with company B's '32mm waste' pipe. Why is this? Well, the '32mm' refers to the internal diameter of the pipe. Sadly, all pipes are joined together via the *external* diameter, and it's this that varies from manufacturer to manufacturer and from one type of plastic to another.

There are three main types of waste pipe: push-fit, compression and solvent weld.

PUSH-FIT WASTE PIPEWORK

A number of manufacturers make push-fit waste pipework and none of them fit with each other at all. Whilst design and details may differ they all use a rubber seal to ensure a watertight connection and, regardless of the manufacturer, they all work the same way, *ie* you take your pipe, you take your fitting and you push the fitting on to the pipe until it won't go any further.

1 The most common error with push-fit waste pipes is pushing the rubber seal off as you force the pipe into the fitting. To reduce the chances of this happening, cut your pipe with a proper pipe-cutting tool rather than a hacksaw.

2 Once the pipe is cut use a file, an abrasive strip or the back of a utility knife to take the edge off the outside of the pipe, creating a slight bevel or chamfer that will slip over the rubber seal far more easily than a sharp, square edge.

3 Finally, to ensure the seal remains intact as you push the pipe into the fitting, it's always a good idea to spray both the pipe and the inside of the fitting with silicone grease.

⚠ **WARNING**

Most rubbers deteriorate in the presence of petroleum-based lubricants, so to ensure a carefree life avoid using WD40 on push-fit waste pipes. If you can't find any silicon spray washing up liquid is a viable alternative, as is plain old water.

Push-fit waste pipe is easy to use and can easily be taken apart again. Sometimes this is a handy feature, sometimes it's not a good idea at all. Over time push-fit connectors can slip off the end of a pipe and create a leak – annoying under your sink, potentially disastrous under your floor. With this in mind you ought to avoid push-fit waste pipe if the pipework is going to be hidden in walls or under floors.

SOLVENT WELD PIPE

Regardless of the size stated, solvent weld pipe won't fit the equivalent sized push-fit pipe; however, most solvent weld fittings will work on most solvent weld pipe. As the name suggests, the waterproof seal here isn't made by a rubber ring but by melting the pipe to the fitting using a solvent.

1 To ensure a waterproof connection you need to cut the pipe using a proper pipe cutter.

2 Now clean it using an abrasive strip.

3 It usually only takes a few seconds for the joint to be absolutely solid and once it is it can't be turned or adjusted. With this in mind the best way to assemble waste pipework is to cut and join everything first, without using glue.

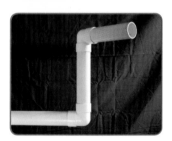

4 Once you have everything cut right and fitting perfectly make pencil marks where the fitting joins its pipework.

5 Repeat this for all the fittings.

6 Now dismantle your creation and get ready to glue it back together.

7 Run a line of solvent around the end of the pipe...

8 ...and around the inside of the fitting.

9 Now push the two together, twisting gently to align your pencil marks.

For some reason the glue makers seem to think it's best if you don't give the fitting a little twist as you put it in. Personally I always do, as I think it smears the glue around a bit more and reduces the chance of any gaps in the seal.

10 Wipe away any excess glue and you're done.

The advantage of solvent weld is that the fittings don't slip out of the pipe, so it's a far more permanent and reliable solution. On the downside it's far more difficult to adjust and you need to do the gluing in a well-ventilated area to avoid inhaling the fumes.

TIP *You may have noticed that whilst the plastic pipework in your bathroom is still fighting-fit after ten years, the same cannot be said for the pipework outdoors. In fact this plastic is now looking distinctly brittle and careworn, with leaks developing here, there and everywhere.*

This effect is called 'UV degradation' and is caused by sunlight slowly breaking down the plastic. Many plastics suffer from this malaise but an exception is a plastic called PVC-C (also known as MUPVC). With this plastic, whilst the colour might fade a little the plastic itself will remain fine and dandy, regardless of what the sun cares to throw at it. Other common plastics used in pipework, such as ABS, counter UV degradation by having a UV stabiliser added to them. However, this tends to just slow down rather than stop the deterioration, but it's better than nothing.

In summary, always check what type of plastic your pipework is made from and verify with the retailer that it's suited to an outdoor life. If in doubt, paint the pipe to protect it from the sun.

COMPRESSION WASTE FITTINGS

So far I've been saying that no waste pipe fits another that isn't identical to it. Well this is where compression fittings come to the rescue. A compression fitting will usually fit any type of pipe of the same nominal size, *ie* a 32mm compression fitting will fit any 32mm waste pipe. For this reason I tend to use only one type of plastic pipework, and use a compression fitting (often called a 'universal fitting') to connect this specific type to whatever I happen to have come across.

To use a compression fitting you first need to take it apart.

1 Undo the nut and slide this on to the pipework.

2 Take out the hard plastic ring and slip this over the pipework.

3 Now push the rubber washer on to the pipe. This is often tapered on only one side. If this is the case this taper should be pointing towards the fitting.

4 Finally, push the fitting on to the end of the pipe and slide everything else down to it.

5 Start turning the nut by hand. Keep in mind that it's very easy to 'thread' plastic fittings by not aligning the nut and fitting properly and then tightening them with a vengeance. To prevent this, apply minimal pressure to tighten the nut until you're certain that it's on properly. If it doesn't go on very far doing this then you've probably not got it on properly, so take it apart and start again.

6 It's usually possible to get a watertight seal just tightening the nut by hand, but if you do get a leak give it an extra tweak using a pair of pump pliers.

CHROME WASTE PIPE

Chrome waste can look lovely and is a delight to those of the frilly-cuff school of bathroom design. Sadly it's not so great from a plumbing point of view. The fittings are usually of the compression type, which is fine except that the rubber seals tend to slip off chrome pipework far more readily. Chrome pipe is also very expensive, so you can't really run all your waste pipe in it, which means that eventually you're going to have to fit it into conventional plastic waste. Sadly, chrome pipework doesn't really fit most plastic pipework so you end up having to use a plastic universal compression fitting to get a good seal.

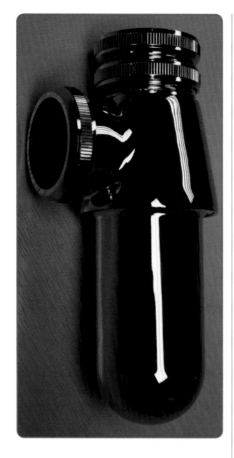

That said, they now do 'chrome-effect' plastic compression joints, which look the part but fit perfectly on to plastic pipe as well.

Name that fitting

One of the biggest problems you may face on entering the wacky world of plumbing is what on earth is everything called. Well, to ease that problem let's have a look at the more common types of fitting:

STRAIGHT COUPLING

Also known as a 'socket', this is used to connect two straight bits of pipe.

SLIP STRAIGHT COUPLING

These look the same as a normal straight coupling but can be slipped over the pipe entirely. When there's no give in your pipework you'll thank the Lord for these. The only issue with using them is making sure they're in the right place, so it's best to mark the pipework first using a pencil.

ELBOW

These are your standard 90° bends.

STREET ELBOW

These are just like a standard elbow except that one end of the elbow slips into a fitting rather than over a pipe. This is really useful when two bends are required very close to each other.

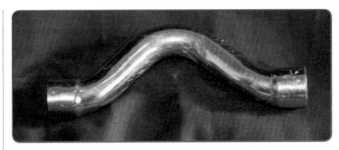

EQUAL TEE

The 'equal' bit refers to the fact that all three connections are the same size, *ie* all 15mm. However, it's possible to buy 'unequal tees', the most common one being where the branch (the side connection) is smaller – the one shown is a 35mm tee with a 22mm branch.

FULL CROSSOVER

See partial crossover.

STOP-END

These are used to cap-off the end of a pipe, either permanently or as a temporary measure.

TAP CONNECTOR

REDUCER

When ordering one, just add to the name what it's reducing from and to, *ie* 'Could I please have a 22–15mm reducer.' You can also buy reducers that fit inside other fittings, such as this push-fit one. The stick-like end fits into a 22mm fitting. The other end is a standard 15mm connection.

PARTIAL CROSSOVER

Only available in copper and regarded as a cheat by professional plumbers but darned handy if you need to take one pipe over another and haven't got a pipe bender.

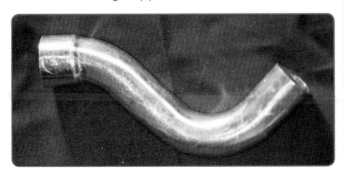

These can be straight, bent or flexi. To make life a tad more awkward tap sizes are normally referred to in imperial measurements, *ie* a basin tap is a ½in tap whereas a bath tap is a ¾in tap, whilst the pipe connection is usually metric. So you might ask for a ¾in to 22mm, push-fit, flexi, tap connector.

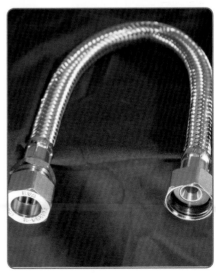

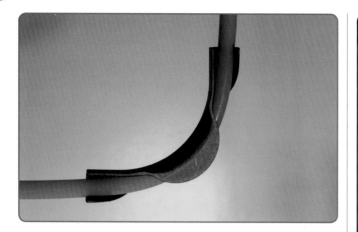

Bending pipework

BENDING PLASTIC PIPE

With plastic push-fit pipework it's often better to put a gentle bend on the pipe rather than use an elbow joint. The advantages of this approach are speed, cost and durability; there's no fitting to play around with, or buy, or leak. The only disadvantage is that you have to make *gentle* bends; if you try to bend the pipe too much it will collapse under the strain and end up blocking the water flow.

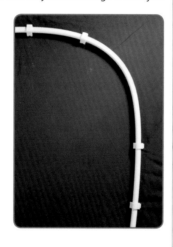

To make it easier to gauge the bend, push-fit systems all come with some sort of bend support, which is just a hard plastic bend that your pipe fits into. However, it's just as easy to create the bend yourself using pipe clips at the start, end and middle to support the bend.

BENDING COPPER PIPE

The advantage of bending copper is that the resultant bend is less restrictive to water flow than a 90° elbow fitting – which is handy for things like power showers. They're also not going to leak, so you can use them underground and in concrete. You can even create bends other than 90°, which is always handy.

To create bends with copper tube you need a specialist tool, cunningly called a pipe bender. Most pipe benders are designed for a number of pipe diameters: the smaller ones usually do 8, 10 and 12mm pipe, whilst the larger are for 15 and 22mm. If you need to bend copper tube wider than this you either need to dedicate a considerable amount of time to gym work, buy an expensive powered bender, or just accept that copper tube above 22mm really doesn't like being bent and use fittings instead.

USING A PIPE BENDER FOR MICROBORE

The smaller benders are used for microbore copper tube. This comes in rolls and can often be bent by hand. However, to get a sharper, cleaner bend you'll need a bending machine.

1 Open the bender fully so the arms are at 180° to each other and slide the copper tube between the arms, making sure it goes under the retaining arm and into the right-sized channel.

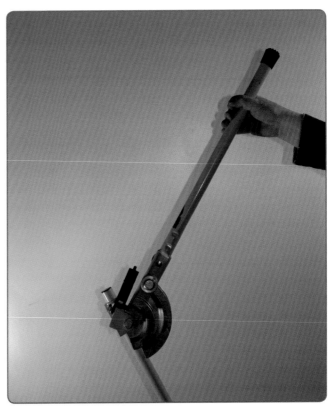

2 Now bend the top arm so that it's parallel with the pipe, which the grooves have now enclosed. Start drawing the top arm down to start the bender.

3 Pull the arms together in one continuous movement until the right angle is reached.

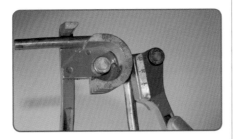

4 Open the arms again to release the bent tube.

5 Pull the tube out of the bender.

USING A PIPE BENDER FOR 15 AND 22MM COPPER

1 Place one arm of the bender on the floor and lift the other as high as it will go.

2 The bender has two 'formers' – one for 15mm pipe and one for 22mm. Place the pipe into the correct former (15mm in this example).

3 Now slide the retaining arm over the pipe to lock it into position.

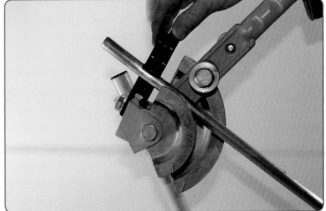

4 Select the right guide for your pipe size and put it on to the pipe, ensuring that the grooved side is on the pipe.

5 With the bottom arm propped on the floor, take the upper arm and start to pull it down.

6 Keep pulling down, trying to maintain an even pressure and constant speed, until you reach the desired angle.

7 Once you've achieved the bend you desire, bend it a little bit more. This is for two reasons: firstly, when you release the pressure the bend will open up a little; and secondly, you can always pull a bend open a little, whereas trying to push the bend a little tighter will invariably result in the pipe collapsing.

8 Open up the arm and release the guide.

9 Pull back the retaining arm and pull the bend out of the bender.

Pipe sizing

The easiest way to truly bugger up the plumbing in your home is to use the wrong-sized pipe in the wrong place. With this in mind, let us look at the most common pipe sizes:

8mm and 10mm TUBE

This usually comes in copper or plastic rolls, and was once used extensively for CH systems (where it was termed 'microbore') because it was cheaper and easier to use. This size of pipe is too small for most applications outside of CH systems, and if you're using it for central heating you should only be using it to feed a single radiator, and not 'daisy chaining' umpteen radiators together.

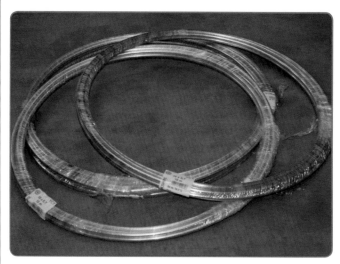

15mm TUBE

The most common pipe size in modern homes. All your mains-fed cold water should be in 15mm pipework, and the hot-water pipework that feeds everything except your bath is usually in this size. Many CH systems are also based around this size of pipework.

22mm and ¾in TUBE

This size is usually only used to deliver gravity-fed water to baths or showers, but it also forms the backbone of your hot-water and CH systems. The cold water entering your home may well be via this size pipework.

TIP *In older houses the pipe size is often ¾in copper, which can no longer be bought. To look at there's no difference between ¾in and 22mm pipework, but when you start to use them together you realise that the ¾in pipe is just a tiny bit smaller, just small enough for 22mm olives not to fit and solder not to run into the joint properly. Fortunately you can buy ¾in olives and use these in a standard 22mm compression fitting. You can also buy ¾in to 22mm straight couplings to convert from ¾in to 22mm. Even better is the fact that the size difference is too small for push-fit plastic fittings to notice, so you can use a standard 22mm push-fit joint on ¾in pipework.*

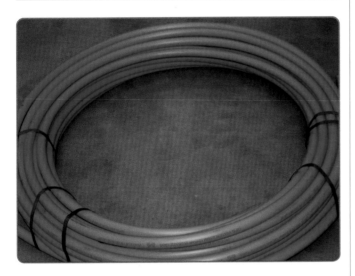

25mm PLASTIC PIPE
This size of pipe delivers the mains water in most modern homes. It's usually blue and terminates at the mains stop tap.

28mm TUBE
Not used very often in modern homes but found in older and larger houses, where it feeds the hot-water cylinder and often forms the backbone of CH systems.

32mm WASTE PIPE
This is the minimum size you should use to remove waste water from washbasins.

40mm WASTE PIPE
This is the minimum size you should use to remove waste water from baths, showers and kitchen sinks.

110mm WASTE PIPE
Also called 4in soil pipe, this is the pipework that your toilet connects to and is the size that forms the backbone of your home's waste system.

Supporting pipework
The two most common causes of noisy pipework are pipes rubbing against something, usually a floorboard, and pipes vibrating because they haven't been clipped into place properly.

Plastic pipe clips are relatively cheap, easy to use and essential if your plumbing is going to remain safe and quiet. There are three types you might come across and use.

The first is just a plastic circle with about a third of the circle cut out. To use these, push the pipe against the circle until it clicks into place.

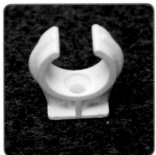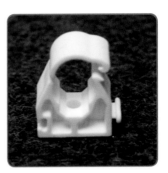

The second is the same thing but with a little arc of plastic that comes over the pipe, clicks into place and locks the pipe into position.

The third type is usually made of metal, has a longer stalk and is usually referred to as a 'stand-off clip'. The point of this type is that, once fitted, there's a large enough gap between the pipe and the wall to fit insulation over the pipe – essential if you're running pipework outdoors.

Pipes rarely run alone for very long, so most pipe-clips are either designed to lock into each other or can be bought already linked. Either way makes life easier and ensures a standard gap between the pipes.

If you don't use enough pipe-clips there's a very real chance that your pipework will vibrate and hum in an astonishingly annoying fashion every time water starts to flow through it. If you use too many pipe-clips you're just wasting a small amount of money. With this in mind, follow the guides below for the advised distances and always add too many rather than too few.

Pipe size (mm)	Copper (metres)		Plastic (metres)	
	horizontal	vertical	horizontal	vertical
15	1.2	1.8	0.6	1.2
22	1.8	2.4	0.7	1.4
28	1.8	2.4	0.8	1.5

7 MAINTAINING AND REPAIRING EMERGENCY VALVES

The water company's stop tap   75

The mains stop tap (or stopcock)   75

The hot-water stop tap   78

Isolation valves   79

In earlier chapters we discussed finding the taps and valves that can save you from a flood in times of emergency. Sadly, all too often these taps don't work, and as a general rule we only discover this mid-emergency. So rather than wade across to the faulty valve and dive down to make a repair à la *Poseidon Adventure* we'll have a look at these emergency valves now, and see how we can maintain and repair them whilst we're still snug and warm.

The water company's stop tap

As we've already mentioned, this is the tap that's outdoors, usually on the boundary of your property. It's underground, hidden beneath a lid, and belongs to the water company. You should never attempt any repair on this tap but it's a good idea to check that it's working (once a year should suffice). If it's not turning off all your cold water completely then call the water company and ask for it to be repaired. They normally give you a date far into the future to do this and then invariably sort it all out within a few days.

For details on how to operate this tap go to 'Where is the outdoor stopcock?' in Chapter 2.

The mains stop tap (or stopcock)

If you've followed the guidance in earlier chapters you ought to know where this tap is. Having found it you need to check every six months or so that it's still working. The test is simple and obvious – turn it off and see what happens. The most common problems are:

- The tap turns freely but it's now leaking a small amount of water from near the head of the tap.
- The water doesn't turn off when the tap is closed.
- You can't actually turn the tap head at all.

REPACKING THE GLAND NUT

Lets deal with the first problem first. If you look at the diagram you can see that the stop tap has what's known as a 'gland nut', and this is generally where the tap leaks – usually if it hasn't been turned for an age. To stop the leak just grab an adjustable spanner and gently turn the nut clockwise until the water stops.

Nine times out of ten this solves the problem, but occasionally

you'll discover that the tap is now very difficult to turn. If this is the case you need to repack the gland nut. This means filling the gland with a waterproof seal, aka PTFE tape.

Tools and materials
- Small adjustable spanner
- PTFE tape
- Small flat-headed screwdriver

1 To be on the safe side it's best to turn off your cold water at the outdoor stop tap and drain down the cold water before you do anything else.

2 With that done you need to turn the gland nut anticlockwise using an adjustable spanner until it comes free and can be pulled back up the tap spindle. Some water may start coming out of the tap as you do this so it's best to put a sponge underneath.

3 You should now be able to see a little gap all around the spindle, and this is where you need to put your PTFE tape. Take the PTFE roll and pull away about 5-10cm of tape – you can always add more tape later but it's a bit of a bugger taking it out if you've added too much, so underdo it rather than overdo it. Now wrap the PTFE tape clockwise around the base of the spindle as close to the gland hole as you can.

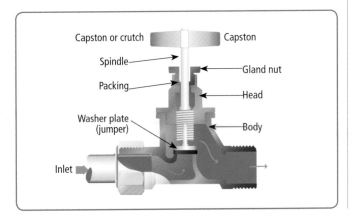

4 Don't wrap it too tight, as once you've finished wrapping it you're going to use a small flat-headed screwdriver to push the PTFE into the hole.

5 With this done, you now push the gland nut back down the spindle and turn it clockwise to tighten it. If you over-tighten the nut you're just going to make it difficult to turn the tap, so just tighten it until you feel some resistance but with the tap head still turning freely.

6 If you now turn your cold water back on you should have a nice dry stop tap that turns freely. If it's still leaking then just turn the gland nut a quarter-turn more and see if that stops it. If not, give it another quarter-turn. If you've put the right amount of packing into the gland you should get to a point where the tap is dry and can still be turned easily. If by the time it's dry you're struggling to turn it, then start the whole process again, adding a little more PTFE.

REPLACING THE WASHER ON THE STOP TAP
If you can turn the tap head but this doesn't stop the water running then the odds are the washer needs replacing.

Tools and materials
- Adjustable spanner or pump pliers
- Rubber tap washers
- Fibre washers
- Small flat-headed screwdriver
- WD40 or a blowtorch (you might need these to get the tap head off) or a proprietary freeze-shock spray

It's also best to have someone available to turn the water on whilst you keep an eye on your handiwork

1 Before you start you need to turn off the cold water at the outdoor stop tap. Once you've done this you need to open up all your cold-water taps to drain the system completely.

2 With this done you're still going to have some water sitting just above the stop tap, and to drain this the stop tap will normally have a drain cock very close to it. There should be no more than a couple of litres remaining in the system so you ought to be able to put a bowl under the drain cock to catch this. If you have doubts about this, or just can't physically fit a bowl under the drain cock, then attach a hosepipe to the end of the drain cock, keep it in place using a jubilee clip and run the hose into a drain.

(If you can't find a drain cock or the one you've found doesn't work, you can still go ahead, but when you remove the tap head you'll get a bit wet, so get ready with a bucket and sponge.)

3 Using an adjustable spanner, open the drain cock by turning the little nut at the end of it anticlockwise until water starts to run out.

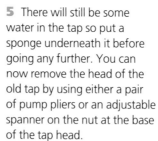

4 When no more water emerges from the drain cock, close it by turning the nut clockwise until you can feel some resistance, then give it another quarter-turn.

5 There will still be some water in the tap so put a sponge underneath it before going any further. You can now remove the head of the old tap by using either a pair of pump pliers or an adjustable spanner on the nut at the base of the tap head.

If the head is on really tightly apply a bit of WD40 and let it soak in for ten minutes before giving it another go. If it's still not moving then either ponder giving up or get a blowtorch or a heat gun and apply heat to the point where the tap head screws on to the body. Once all is really hot try to turn it again. If it still isn't happening then freeze-shock it using a spray, if this fails give up and replace the whole tap (see next page).

6 With the old head removed you can now see the washer. It's not unknown for this to have disintegrated completely, so don't be too surprised if you can't find it.

This is also a good time to check the overall state of the tap head. If it's looking pretty poor then consider replacing it altogether. This means buying a new 15mm stop tap, removing the head from it and screwing it on to the old tap body. For very old taps this isn't always an option but nine times out of ten this approach works perfectly well.

The old washer can be prised off by hand, or you can use a flat-headed screwdriver to make things easier.

7 Find a replacement washer. A washer that's a little smaller than the original will usually work fine, but one that's a little larger will invariably get stuck in the tap body and fall off.

8 The tap head should have a fibre washer on it to ensure a waterproof seal when you tighten it back on to the body. Check that this is in a decent state. If it isn't remove it and fit a new one. On some new taps this washer is made from hard plastic.

9 Fit the tap head back on and hand-tighten it to the body. Once hand-tight use the adjustable spanner to give it an extra quarter to half turn.

10 With that done, check that all your cold-water taps are closed, double-check that the drain cock is closed and that the new stop tap head is closed. Now turn on the water outside and check that all remains dry as you open the stop tap and let the cold water back into your home – it's always a good idea to have a partner for this process so one of you can open the outside tap and the other can stay indoors and scream if a leak appears.

I CAN'T TURN THE TAP HEAD AT ALL!

It's not unknown for the tap head to seize altogether. There are ways of getting around this:

- Loosen the gland nut and see if this loosens the head enough to turn it. If this works you'll probably still need to repack the gland to stop any leaks – see above.
- Replace the entire tap head. To do this just buy a new 15mm stop tap and follow the instructions for re-washering the tap (see above), and fit the new head on to the old body. This doesn't always work but usually does. If you're out of luck and you can't get the new head to fit the old body, replace the entire tap by following the instructions below.

REPLACING A STOP TAP

Tools and materials
- Pump pliers
- Adjustable spanner
- Jointing compound suitable for potable water, or some PTFE tape

1 The first thing you need to check is the size of the pipes coming into and leaving the stop tap. As a general rule the pipes leaving the tap will be 15 or 22mm copper, while the pipes coming into the tap might be:

- 15mm or 22mm copper
- Blue plastic – usually 20mm or 25mm
- Black plastic – usually ½in imperial
- Lead

Ideally you want to jot down what the pipe sizes coming into the old tap are and then pop down to the plumbing store and buy a replacement. However, you might actually find it easier to take the old tap out and pop into the plumbing store and ask for 'one of these.' The obvious risk in this approach, though, is that no one has the replacement tap in stock.

2 Having purchased a new tap, turn off the cold water using the outside stop tap and drain the cold water down as described earlier. Remove the old tap by holding it steady with a set of pump pliers and turning the nuts at the top and bottom with an adjustable spanner.

3 With the tap out you can usually just leave the old nuts and olives in place, unless they're what was leaking, and fit the new tap to these. If you're going to opt for this approach it's best to put a little jointing compound around the olives first – ensuring that you use a compound that's suitable for drinking water. If you decide to remove the nuts and olives follow the guidance in 'Removing compression fittings' (Chapter 6).

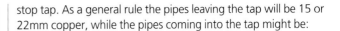

4 Fit the new tap into place, bearing in mind that most have an arrow on them indicating how they expect the water to flow. Tighten the old nuts on to it. Hand-tighten and then give them a three-quarter turn with the adjustable spanner – you might need to hold the tap in place with a set of pump pliers whilst you do this.

5 Turn the new tap off. Make sure all drain cocks and taps are closed and turn your water back on and check for leaks.

The hot-water stop tap

If you have a gravity hot-water system, or have an unvented cylinder, then this is the tap close to the cylinder itself that turns off the hot water to your home. This tap will usually be of a type called a gate valve. It works by turning the head (usually red) clockwise. This in turn lowers a metal bar (or gate) that stops the flow of water. The advantage of this type of valve is that when open it doesn't restrict the flow of water. The disadvantage is that the gate tends to get corroded in the open position. When this happens the temptation is to apply more and more pressure to the head in order to free it. Sometimes this works, but often the spindle connecting the head to the gate just breaks and you're left with nothing more than a useless lump of metal affixed to your pipework.

To stop this happening it's important to open and close this type of valve on a regular basis – twice a year at least. Sadly few people actually do this, so sooner or later the valve just corrodes away and needs replacing.

REPLACING THE HOT-WATER STOP TAP

Tools and materials
■ New stop tap or lever valve
■ Pump pliers
■ Adjustable spanner
■ Jointing compound (suitable for potable water) or PTFE tape

As already mentioned, the important aspect of a gate valve is that it doesn't restrict water flow. Consequently you shouldn't replace this type of valve with any old tap; you should always replace it with either an identical gate valve or with a full-bore lever valve. Personally I always replace gate valves with lever valves, because these tend to work for longer, even when they're neglected for years on end, and are easier to turn.

1 To remove the hot-water stop tap you first need to drain the system. To do this, make sure the hot-water stop tap is in the open position, then turn off the cold water at the mains stop tap and open all the hot-water taps in the house. At this point the cold-water storage tank up in the loft will start to empty, and gradually the water coming out of the hot-water taps will turn cold and then cease altogether. At this stage you'll have emptied the cold-water storage tank but will still have a full hot-water cylinder.

If you have an unvented cylinder then just ignore the tank in the diagram below. In all other respects what happens to the water is the same, *ie* the cylinder itself is still full.

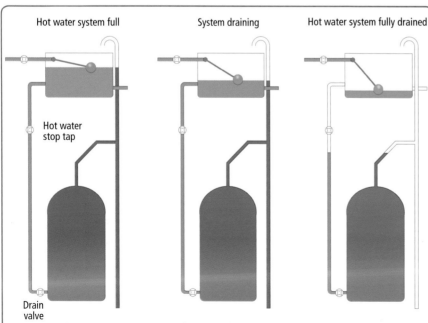

Hot water system full   System draining   Hot water system fully drained

Hot water stop tap

Drain valve

2 Carefully loosen the bottom nut on the gate valve. You shouldn't get more than a dribble of water appearing, so catch that with a sponge and pull the pipe out of the fitting.

3 Have a quick look into to the fitting if you can, just to confirm that the valve is open, before loosening the top nut and removing the valve completely.

Gate half closed

If you're going to replace the old valve with an identical gate valve you can usually reuse the old nuts and olives – just apply a small amount of jointing compound (suitable for potable water) to the olives first.

If you're going to replace the old gate valve with a full-bore lever valve you might find that the old olives and nuts don't quite fit – often the olives are too far down the pipe. If this is the case, take off the old olives (see 'Removing compression fittings' in Chapter 6) before fitting the new lever valve.

TIP *Always aim to fit lever valves so that if the lever arm drops accidentally it will close the valve rather than open it, ie arrange it so that it's fail-safe.*

Open

Dropping to closed position

4 Fit the new valve and tighten the nuts on either end using an adjustable spanner – read the section on compression fittings in Chapter 6 if you have any doubts.

REFILLING YOUR HOT-WATER SYSTEM

For an unvented cylinder, refilling the system is just a matter of turning the cold water back on whilst ensuring that the hot-water tap farthest from the cylinder is open. When water starts to come out of this tap turn it off, and that's it!

With a gravity-fed system things are more awkward because there's a real danger of getting an airlock when you refill it. To minimise the chances of this follow these steps:

1 Close the new valve you've fitted and open up the mains cold water again. The cold-water storage tank will start to fill. Whilst it's doing that, check that all the hot-water taps you opened to drain the system are still open.

2 When the cold-water storage tank is almost full open the new hot-water stop tap. At this point you should hear a number of gurgles and water should start to emerge from the hot-water taps downstairs. Initially the water will come out in fits and starts as the air in the pipes is pushed out, but eventually the flow will steady.

3 Once the flow is steady turn that tap off and check the next one downstairs. When you've closed all the downstairs hot taps check the upstairs ones until you have a smooth flow of hot water from all the taps.

If you still get an airlock refer to 'Dealing with an airlock' in Chapter 8.

Isolation valves

The idea behind an isolation valve is simple: if a tap is dripping, or a toilet cistern is overflowing, it's much more convenient to turn

off that single tap or single cistern than have to turn off the cold or hot water to the entire house.

An isolation valve is what's known as a sphere valve, and is essentially a valve with a ball bearing in it; but here's the clever bit – they've drilled a hole right through the ball bearing! When you rotate the ball bearing so that the hole aligns with the pipe the valve is open and water flows through the pipe, through the hole and out the other side. Turn the ball bearing so that the hole is at right angles to the pipe and the pipe's completely blocked. The advantages of this approach are that there are no washers to disintegrate and split over time, and isolation valves are much more resistant to the effects of limescale.

To allow you to rotate this ball bearing the valve usually comes with a little groove in it. You put a flat-headed screwdriver into this groove and turn it so it's either aligned with the pipe (open) or turned at 90° (closed).

Open — Closed

If the ball bearing is rotated by a lever the valve is usually called a lever valve. The lever makes these much easier to open and close, but they're not as visually appealing, so they're usually only used when the valve is hidden away in a loft or cupboard. Lever valves designed so that the hole in the ball bearing is the same diameter as the pipework are called full-bore lever valves. These are essential when isolating things like showers, where you don't want to inhibit the flow of water.

Isolation valves come in compression and push-fit forms, the latter often being made entirely of plastic. When this is the case the groove for the screwdriver is often replaced with a stubby handle. Occasionally you might come across a push-fit isolation valve where the groove is also made from plastic, but since the groove gets worn away very quickly by a screwdriver this type is best avoided.

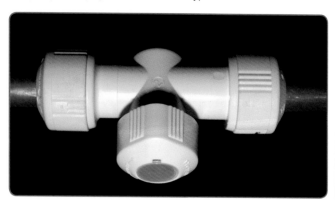

WHERE TO PUT ISOLATION VALVES

The water regulations say that all new float-operated valves should have an isolation valve very close to them. This means that your toilet cistern and the storage tanks in your loft should all have isolation valves fitted to the cold-water pipework, just before the float valve – also called a ball valve. The reason for this is quite simple: float valves fail, and when they fail they tend to fail badly, *ie* water gushes out of them and just doesn't stop. So when this happens it's very handy to be able to close an isolation valve and stop a possible flood and a definite wastage of water.

So why don't you have isolation valves already fitted? Because it's only a MUST for new installations, and if yours was fitted before the regulations came into force you don't have to retro-fit one now. However, the sound reasons for fitting them don't change; they stop you wasting water and they make life easier, so fit them anyway.

The other eminently sensible place for an isolation valve is before all the taps in your home, as well as any outside taps. Your shower should also have isolation valves, but in this instance you should fit full-bore lever valves so as not to restrict the flow of water.

FITTING AN ISOLATION VALVE

Tools and materials
- Adjustable spanner
- Pump pliers
- Abrasive strip
- Utility knife

1 The first step is to decide on a location. You really want to get them as close to the tap or float valve as is sensibly possible, whilst ensuring that you can easily turn them on and off. It's also essential that the pipework is straight and clean. If the location is on view, consider using a chrome valve in either compression or push-fit form, as these are quite pleasing on the eye.

2 Having decided on the location, turn off the water and drain the system.

3 Cut into the pipe using an appropriate pipe cutter.

4 The isolation valve itself uses a bit of space, so you'll need to cut a section out of the pipe to accommodate this. How deep the pipework fits into the isolation valve is usually marked on the valve, so you can just measure between these marks to see how much of a gap you need to make, but as a general rule the gap is the same diameter as the pipework itself, usually 15 or 22mm. Remember that there's usually some

movement in the pipework, so you can always adjust this gap if necessary.

5 To ensure that the pipework is clean and free of paint and debris, take an abrasive strip to the pipework and apply a little elbow grease – don't overdo this on plastic pipe.

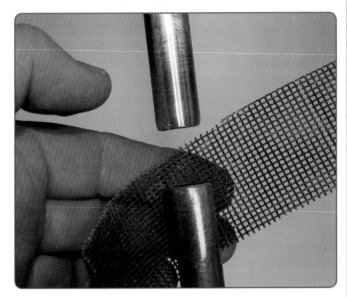

TIP *Removing paint from copper pipe is a real chore but is absolutely essential if you desire a leak-free life. There are several ways you can go about it but my favourite is to open up a set of pump pliers, gently grip the pipe with them and then turn the pliers to and fro a few times. This usually causes the paint to flake off and you can then finish the job with an abrasive strip. Alternatively you can try scraping the paint off with the back of a knife or heating it up with a blowtorch. Whatever approach you use, always finish off by using an abrasive strip to get it nice and smooth and shiny.*

6 With the exception of the full-bore type most isolation valves have an arrow on them to indicate the direction that water should flow through them. To be honest they seem to work fine the wrong way around, but apparently they can seep water, so always check for this arrow and fit the valve accordingly.

7 With the pipe clean you're ready to fit the valve. You can dismantle a compression-type valve first if you prefer (see Chapter 6), but it's usually fine to push clean pipe straight into a compression joint – just check that the olives are aligned properly first.

8 Finger-tighten the nut, then give it a three-quarter turn with an adjustable spanner, using a second spanner to keep the valve steady whilst you tighten.

If you're using a push-fit valve ensure that the pipe is fully inserted and try your level best to push the pipe in straight rather than at an angle, to avoid damaging the rubber seal.

9 With the valve fitted use a flat-headed screwdriver to close it. Then turn the water back on.

10 Check for any leaks by running your finger around the base of the valve. If all is dry open the valve, let some water run out of the tap or ball valve, then close the tap/ball valve and check that there are still no leaks.

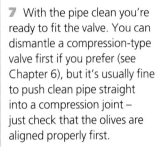

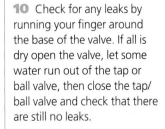

8 BASIC PLUMBING: THE USUAL SUSPECTS

Help, I have a leak!	83
Leaking compression joints	83
Leaking washers	84
My sink is blocked	85
My toilet is blocked	88
Blocked drains	89
My toilet won't flush properly	91
My toilet is overflowing	97
The siphon is leaking	100
The tanks in the loft are overflowing	101
My taps are dripping	102
Dealing with an airlock	107
Fitting a new cold-water storage tank	108

A plumber's life is both wondrous and varied, each day a veritable smorgasbord of unique fascination ... and yet, at the same time, we get an awful lot of the same thing. So let's look at the more common problems, which by happy chance are often the easiest to resolve.

Help, I have a leak!

DETECTING A LEAK

This sounds obvious, and it is, but it's still something that people approach in the most haphazard of manners with varying degrees of success; so let's go through the basic steps.

We're going to assume at this point that little or no guidance is required to detect a jet of water that's emerging from your mains pipe and bouncing off the ceiling. Here we're talking about the subtle leak, that insidious trickle that can cause so much damage simply because it can go unnoticed for so long.

Sometimes these can be so subtle that you're not even sure if you actually have a leak; it could just be condensation on the pipe or fitting. To be honest it can be very difficult to tell the difference, but there are usually a few clues.

Firstly, condensation normally forms on cold pipework in a cold place, so if you're looking at a warm pipe and feeling pretty warm yourself it's almost certainly a leak you're looking at.

Secondly, a leak arises from a specific point, whereas condensation arises when damp air meets a cold pipe. So, to test the difference, dry the pipe off with some tissue and keep an eye on it. If the dampness returns along the whole pipe at pretty much the same time then it's probably condensation. If the dampness seems to start at the top and work its way down then it's probably a leak.

To detect where the leak is coming from follow the dampness on the pipe upwards by running your finger around the pipe and checking to see if your finger is wet – if it's a cold-water pipe your finger will probably feel wet without actually showing any signs of dampness, so actually examine your finger.

Dry your finger on some tissue after each check and work your way up the pipework, paying special attention to the joints – joints are far more likely to leak than the pipework itself.

At some point the pipework should suddenly become dry, meaning you've passed the leak. So work your way back until you find the dampness again. To test your discovery dry the joint and watch as the dampness seeps back.

We've already talked about burst pipes (Chapter 4), but these are rather rare events. The leaks that you're far most likely to discover are those caused by compression joints failing and washers disintegrating.

Leaking compression joints

The problem with compression joints is that after holding back the water for years they'll take one knock too many and that's enough to start them leaking. Most of the time all that's required to stop the leak is to tighten them up a little.

TIGHTENING A COMPRESSION JOINT

Tools and materials
- ■ Pump pliers
- ■ Adjustable spanner
- ■ Tissue paper

1 Hold the joint steady with a set of pump pliers or an adjustable spanner.

2 Put an adjustable spanner on the leaking nut and give it a clockwise turn. It often only takes the tiniest of turns to stop the leak.

3 Use a tissue to dry the joint and then wipe your finger around the nut to check it's dry. If it is, wait five minutes and then wipe your finger around it again. If it's still dry it's safe to say that you're stopped the leak; if it's damp again then give the nut another tweak.

REPAIRING A COMPRESSION JOINT

Tools and materials
- ■ Pump pliers
- ■ Adjustable spanner
- ■ Sponge, bowl or water vacuum
- ■ Jointing compound (suitable for potable water) or PTFE tape
- ■ Tissue paper

Sadly, tightening the nut doesn't always stop the leak; in fact in some cases the leak gets bigger. If this is the case you'll need to drain the water out of the appropriate pipe and repair the joint as follows:

1 Undo the nut by holding the joint steady with one adjustable spanner whilst turning the nut anticlockwise with another. At this point water may start to come out of the joint, so be prepared to catch it with a sponge or something similarly absorbent.

2 Pull the nut back to expose the olive.

3 Pull the pipe back a little to release the other side of the olive from the joint.

4 Apply either a small amount of jointing compound (making sure it's appropriate for drinking water) or take about 10–15cm of PTFE tape and wrap this clockwise around the olive, covering both sides of the olive.

5 Push the pipe back into the joint and finger-tighten the nut.

6 Hold the joint steady and tighten the nut by turning it an additional three-quarter to full turn with an adjustable spanner.

7 Turn the water back on and dry the joint with some tissue paper. Check it's dry with your finger, then wait five minutes and check again. If it's still leaking either the olive has been over-tightened at some point, the olive is damaged, or the pipework is damaged. If this is the case you'll either have to replace the olive or replace the entire section of pipe (see Chapter 6).

TIP *For hot-water pipes and CH pipework always check the joint remains dry once the pipework is hot – it's common for a joint to be perfectly dry when the water is cold, yet leak like a sieve when the water heats up!*

Leaking washers

Washers are usually made from rubber or fibre. Alas, both materials corrode over time. Often this doesn't cause any problems until the pipe is knocked and the washer finally falls apart and starts letting water through.

Tools and materials
- ■ Adjustable spanner or basin spanner
- ■ Small flat-headed screwdriver
- ■ Replacement washer

1 The most common place to find a washer is in the connector at the base of a tap or a ball valve. The first step in resolving the problem is to turn the water off and drain the pipework.

2 Using an adjustable spanner, loosen the nut by turning it anticlockwise. If you're trying this with a tap you may well need a basin spanner.

3 Now pull the nut back and ease the pipework out of the fitting.

4 Remove the old washer. It's not unknown for people to just put a new washer over the old, and doing this will stop the leak – in the short term. In reality adding a second washer is just adding to the problem, so make sure you get every bit of every washer out of the fitting, using a small screwdriver to aid this process if necessary.

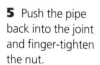

5 Once you're sure the joint is washer-less, put the new washer in place.

6 Push the pipe back into the fitting, making sure that the washer is snugly in place.

7 Push the nut back up and hand-tighten it before using an adjustable spanner – or a basin spanner – to give it another three-quarter turn.

Note that in an emergency you can create a temporary washer by wrapping PTFE tape around the end of the pipe where the washer was. However, this rarely lasts very long, so only employ this approach as a stop-gap whilst you find a proper washer.

TIP *A common scenario is to find a plastic fitting, with a plastic thread (the tails of toilet float valves are notorious for this), fitted to copper pipework with a brass nut. The problem with this arrangement is that the brass nut will slice*

through the plastic thread with almost no encouragement whatsoever, ruining the thread and guaranteeing a leak. To stop this happening always tighten up the nut by hand using the lightest of touches. If the nut is hard to turn you're probably 'threading' the joint, ie the brass is carving its own route through the plastic rather than following the thread itself. If you suspect this, undo the nut, check the pipe is going straight into the fitting and not at a slight angle, check the plastic thread is undamaged, and then start again.

If possible try to have the fitting itself quite loose so that you can be certain that the pipe and fitting are in perfect alignment – this can't be done with taps but is the best approach with ball valves.

My sink is blocked

Prevention, they say, is better than cure, so let's start off by looking at the most common cause of sink blockages, which is fat – fatty food remains off plates, or fat poured down the drain straight out of the chip pan. Please don't do this. Not only does it block your household pipework but it can also block miles and miles of main sewerage pipe.

If you have a pan full of fat then for the sake of your waste pipework, the nation's sewers and your own peace of mind, pour it into a container and throw the lot into the bin when the container's full.

That said, despite your best endeavours sooner or later your sink, washbasin or bath is likely to become blocked. Usually the blockage will be in a bend, and the most obvious series of bends in a waste pipe are in the 'trap' itself, so we'll be looking at this directly a little later. However, let's start off with the obvious approaches to unblocking a drain:

USING A PLUNGER

Tools and materials
- Plunger
- Damp cloth or old towel

1 Plungers often work best when you're firing water down the blockage, so if the sink doesn't already have water in it add some.

2 Put the plunger over the plug outlet.

3 Place a damp cloth over the overflow outlet to close it off – you might need to apply a fair amount of pressure.

4 Rapidly force the plunger down and repeat this action several times.

5 Remove the plunger and see if the water drains away. If it doesn't try repeating steps 2 to 4.

TIP *One of the most impressive ways of shifting a blockage in pipework is a water vacuum cleaner set to blow. You need to ensure all overflows are covered and that plugs are fully closed. Even then the vacuum cleaner will try it's best to blow the plugs across the room.*

On solvent weld pipe this approach can work a treat, but it's not recommended for push-fit pipework as it can blow the joints off the end of the pipes.

If you do opt for this approach bear in mind that whilst it will often deliver instant results, you might have to spend the rest of the day clearing up the mess.

USING A DRAIN CLEANER

The power and effectiveness of drain cleaners has changed over the years in indirect proportion to their safety, *ie* they're not as effective these days but you're far more likely to emerge unscathed after using them. That said, most still contain some fairly violent chemicals, so always wear eye protection and rubber gloves, and follow the manufacturer's instructions to the letter.

Rather than wait until your sink is totally blocked it makes sense to use drain cleaners on a regular basis to stop blockages in the first place.

CLEANING A BOTTLE TRAP

Most washbasins these days tend to be fitted with a bottle trap. Whilst these look neater than traditional traps they do tend to block easier, so avoid their use with kitchen sinks.

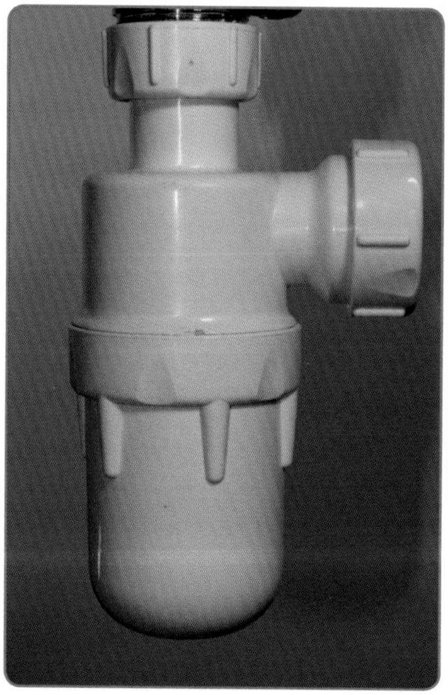

Tools and materials

- Bowl
- Pump pliers
- Screwdriver (not your best one)

1 If possible place a bowl under the trap to catch the water. You might also want to don a nose-plug because this can all get very smelly very quickly – opening a few windows would also be a good idea.

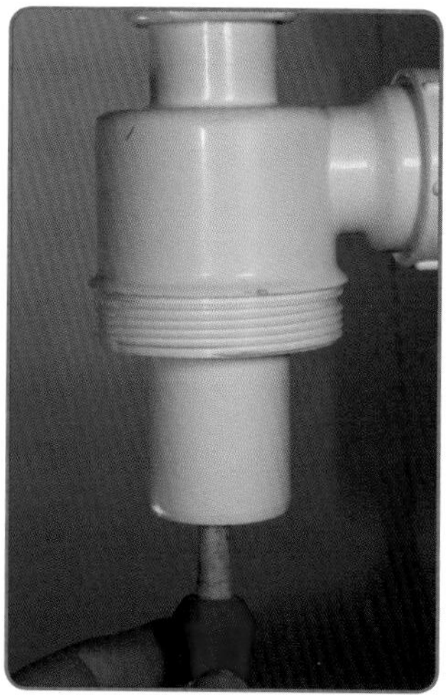

2 Remove the base of the trap either by hand or by using a large pair of pump pliers.

3 Take the base to an outside tap and give it a thorough clean-out.

4 Wash down the inner tube of the trap and check the inside. Use a screwdriver to get into the more awkward areas, and disconnect the bottle trap from the base of the washbasin if it's going to make things easier.

5 Put the trap back together and run the washbasin taps to test all is still dry.

CLEANING AN ORDINARY TRAP

There are two main types of 'ordinary' trap: the S-trap and the P-trap, so-named because their shapes resemble those letters. They often come with an access point that can be used to clean them out. Personally I'd rather just dismantle the whole thing and do a thorough job.

Tools and materials

- Bowl
- Pump pliers
- Bottle brush or screwdriver (not your best one) and some cloth

1 If possible put a bowl or sponge under the trap to catch any water.

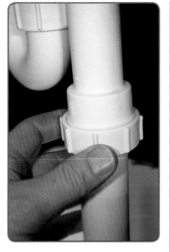

2 Undo the compression nut below the sink waste.

3 Undo the compression joint where the trap meets the waste pipework.

4 Carefully remove the trap and pour the water into the bowl – this will stink!

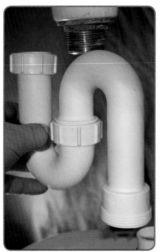

5 Take everything outside and dismantle the trap in its entirety, laying the pieces out on a cloth so that you can remember how to put it back together again. Some traps can be dismantled, some can't. It's pretty obvious how much they can be dismantled, so just undo what nuts you can.

6 Clean out using the outside tap and a bottle brush or a combination of screwdriver, cloth and elbow grease.

7 When you're sure everything is clean put all the bits back together again, checking that all the rubber seals are intact and in place.

8 Fit it back under the sink. Push the top of the trap up against the base of the sink waste so that the rubber seal is in place, then hand-tighten the nut. This is usually sufficient, but give

it a tweak with your pump pliers if you're at all worried.

9 Check the rubber seal on the waste pipe is in place and push the pipe into the trap before tightening the nut by hand. Again, you can give it a little tweak with a set of pump pliers if need be.

10 Run the taps and check all is dry. If not, tighten the nuts a fraction using your pump pliers.

DISMANTLING WASTE PIPEWORK

With really bad blockages you might have to take apart the entire waste pipework. We've discussed waste pipework in Chapter 6, so identify what type of pipe you have (solvent weld or push-fit) and be prepared to either pull the pipework apart or cut into it using a proprietary pipe cutter. Either way it will stink and it will be messy.

Remember, the blockage will almost certainly be at a bend so check these areas first.

My toilet is blocked

If you have a small child in the house then it's not unknown for the toilet to be used as a convenient repository for pretty much anything, from jewellery to small toys. These then get trapped in the toilet U-bend and cause blockages. You might actually want to retrieve the items rather than just clear the blockage, so we'll look at how to remove a toilet in a minute. However, there are times when you just need to unblock the toilet, so let's have a look at other approaches first.

First off, since there's a large risk of getting covered with stuff you'd much rather you didn't get covered with, don old clothes and large rubber gloves before you do anything else.

USING A PLUNGER

1 The plunger shown operates on much the same principle as a bike pump, but to operate it effectively you need to be drawing up

and expelling water rather than air. Push the plunger into the base of the toilet so that it's covering the exit from the pan.

2 Now push it down rapidly, let it come back slowly and repeat the downward thrust.

3 The water level should start to drop once the blockage is cleared. If you think that's the case then remove the plunger and give the toilet a flush.

If you haven't got a plunger handy try using a toilet brush, which often works just as well.

USING A DRAIN AUGER

This is a length of flexible wire that you can rotate and wiggle around, the idea being that this will clear anything around the wire. It's often called a 'drain snake' for fairly obvious reasons.

1 To use it on a toilet push it into the pan and try to push it around the U-bend – sometimes this works, sometimes it doesn't; it rather depends on what type of toilet you have and whether your luck is in or not.

2 Once you have it in place turn the handle and you'll usually hear the wire turning in the pipe.

3 A sudden drop in the level of the water indicates that you've cleared the blockage, so pull the auger out and test the toilet by flushing it.

TAKING A TOILET OUT

What type of toilet you have will determine if removing it to clear a blockage or retrieve whatever your cherished child has hidden in the U-bend is going to be feasible or not.

Most modern (ie white) toilets have a horizontal outlet pipe connected to a plastic pan connector. The pan connector itself can continue horizontally and disappear into the wall behind the toilet, or it can be angled and disappear into the floor. This type of toilet can be removed with relative ease.

Older toilets, aside from being in a wide range of dubious colours, can still have a horizontal outlet but this is connected to a

solid pipe which is usually cemented into place.

Alternatively the toilet outlet goes down into the floor, usually terminating in a solid pipe surrounded by cement. If your

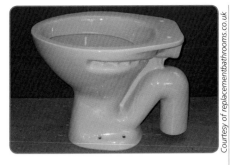

Courtesy of replacementbathrooms.co.uk

toilet looks anything like these two examples then it is almost certainly cemented into the floor and if you try to remove it the odds are it will break. So don't even bother unless you're happy to buy and fit a replacement toilet.

Read 'Removing an old toilet' and 'Fitting a new toilet' in Chapter 11 for the relevant detailed steps.

Blocked drains

If your toilet is backing up then the problem is often in your main drain rather than the toilet. The easy way of checking this is to see if the bath and washbasin etc are also affected. If so, then it's almost certainly a main drain problem.

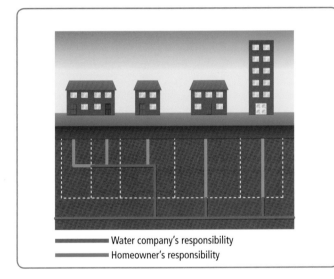

Water company's responsibility
Homeowner's responsibility

From October 2010 you're responsible for those drains that serve only your home and are within your property boundary. Once the drain crosses your property boundary and/or links to a shared drain the water company becomes responsible for it. If you live in a block of flats then the water company doesn't take responsibility for the drain until is crosses the property boundary of the block.

There are two main types of drain you'll come across. One is an open gully and usually

serves the kitchen sink waste. The other is the main underground drain, usually found under a manhole.

OPEN GULLY

This is usually blocked by leaves and other debris and the best way of clearing it is to get down and dirty, ie don a really big pair of rubber gloves, put your hands into the gully and start lifting out all the lovely gunk. Once you've cleared most of it you can try using a drain cleaner to shift the rest (always follow the manufacturer's instructions) or try a drain auger.

Having cleared the gully, stop it getting blocked again by putting a drain cover over it. You can buy these from most DIY stores or make your own from whatever comes to hand.

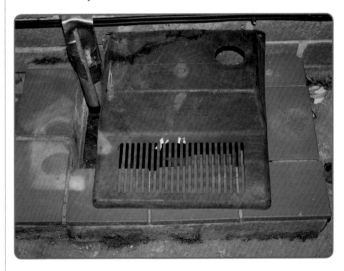

MAIN DRAIN COVER

If you have a blockage you can usually tell roughly where it is by the fact that stuff is seeping out from under the manhole cover. It's best not to enquire too closely as to what this stuff actually is.

Taking off the manhole cover can be a task in itself. There's a little tool for this but the odds of you having one are very slim, so opt for a combination of claw-hammer,

crowbar and the larger and more robust flat-head screwdrivers in your collection.

You might also want to phone a friend for a little help. Even then, removing manhole covers is an ideal opportunity to crush your fingers, so take care and remember that this isn't a race.

With the cover removed a clear drain might reveal a plastic or ceramic half-pipe with connections entering and leaving from various directions. If the drain is blocked removing the cover should reveal something that looks like it's just come out of a B-movie entitled *Day of the Killer Gunk*.

What you need to find is the exact location of the blockage, and the way to do this is to get an idea of which way the waste is moving and then open the drain covers along the route until you find a drain that's clear. Once you've found that you can be pretty sure that the blockage is between the manhole filled with 'stuff' and the one that's clear.

Remember, if the blockage isn't on your property but is affecting your home, it's the responsibility of the water company to unblock the drain.

1 Make sure you have a pair of thick rubber gloves and are wearing old clothes. Now screw two or more poles together and start pushing them down the drain. It can take a while to find the actual drain in the murky depths, so be patient.

2 Add more poles as required and carry on pushing them along the drain. As you push ALWAYS give the poles a clockwise twist. This ensures that they don't unscrew and fall off in the middle of the drain, which is dreadfully important as there's nothing more embarrassing than having to admit that you've just blocked a drain with your drain-clearing kit.

3 Push the poles back and forth as you twist them clockwise and keep adding more lengths. With a little luck this will be enough to clear the drain and you'll see a sudden burst of activity as the gunk starts to gurgle and the level suddenly starts to drop. It's not unusual for it to immediately block itself again, so carry on agitating everything until the level has dropped down to normal and you can see the half-pipes and the drain connections below.

4 Remove your poles by pulling them as you twist clockwise, and unscrew them off the end as they emerge.

5 Grab a hosepipe and spray down your rodding kit and the walls of the manhole. Then run a quantity of clear water down the drain to make sure you really have cleared it.

TIP *Health and safety issues abound in this area, so if you're lifting umpteen drain covers you need to either stay close by them whilst they're open to ensure no one falls in, or put a barrier around them. In other words, use a bit of common sense.*

Now you know where the blockage is you need a rodding kit. This is a series of plastic poles that screw into each other and to which you can attach a variety of interesting appendages. It's best to start off using just the poles, without any fancy bits on the end.

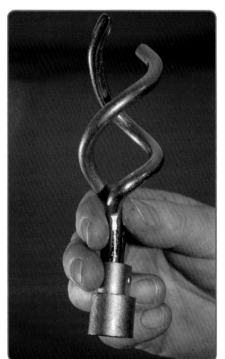

6 If you get no joy using just the poles, then screw one of the appendages on top of the first pole. I always opt for the screw-like device first and the round rubber seal as a last desperate option.

My toilet won't flush properly

The device used to flush a toilet is called a siphon. In olden days this was just a plastic U-bend containing a thin sheet of plastic. As you turned the toilet flush handle the plastic sheet was raised, lifting the water over the top of the U-bend. Once you released the handle, siphonage would cause the rest of the water in the cistern to follow and the toilet was flushed.

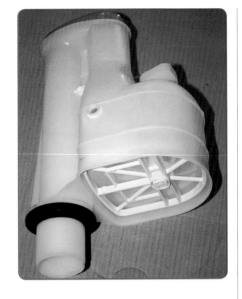

Since those halcyon days of yore there's been some sort of race to create the world's most complicated toilet siphon, with the result that these days you have little clue what you might find when you raise the cistern lid – and most of them work slightly differently.

That said, whilst new siphons might leak (see later in this chapter) they rarely stop flushing. With this in mind, what might cause a toilet to stop flushing? Well, it's going to be one of the following:

- The siphon is broken.
- The handle, or the push buttons, are broken and are no longer operating the siphon mechanism.
- The water level is set wrong and there's not enough water in the cistern to generate a full flush.

We'll look at these in order of simplicity, which means checking the water level first.

SETTING THE CISTERN WATER LEVEL
Carefully take the lid off the cistern, bearing in mind that it might be attached to the flush mechanism.

TIP *With push-button systems the cistern lid is often connected to the flush mechanism. In this case pulling off the lid will break the siphon underneath. So start by gently trying to lift the lid. If you feel resistance, stop and have a look at the button mechanism.*

Each design is different, so if possible find out what type you have and either find the manual or look it up online. That said, the following basic steps work in most circumstances.

There are usually two buttons (half and full flush) and you can usually remove these by depressing one and then putting a small screwdriver under the other and lifting it out – make a note of which button is where, eg small button on right, large button on left.

Once you've removed the buttons look inside the mechanism. If you can see a large plastic screw-head use a flat-headed screwdriver to loosen this off and the button mechanism should then come away. If there's no screw the odds are the entire button mechanism is screwed into the cistern lid. If this is the case turn the mechanism anticlockwise to loosen and remove it.

1 With the lid removed, check the water level. There's usually a mark set into the cistern indicating what it should be, though this isn't always easy to see. If the water level is at that mark then this isn't the cause of the flush not working. If the water level is well below this mark then you need to adjust your ball valve.

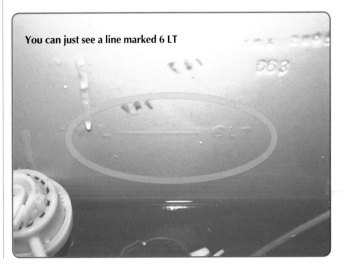

You can just see a line marked 6 LT

2 In older models you adjust the water level by loosening off the ball valve and just moving it slightly higher up the arm. In really old models you have to resort to bending the arm up or down.

In slightly newer

models you adjust the height by turning the screw at the base of the arm – you might need to loosen the nut first.

In yet newer models you usually adjust the level by turning a plastic screw, which moves the float up and down. Turn it clockwise to raise the water level, anticlockwise to lower it.

3 Flush the toilet and see what the level is now. If all is OK just put the cistern lid back on. Otherwise, repeat these steps until you have the level right.

REPLACING A BROKEN TOILET HANDLE

The inside of a toilet cistern is a cold, damp place and toilet handles can slowly rust away until the spindle or arm breaks. By lifting the lid and operating the handle you can usually tell very quickly if the handle is actually working or not.

Tools and materials
- Pozi screwdriver
- Hacksaw
- Adjustable spanner

1 Carefully take off the cistern lid.

2 Use a screwdriver to loosen the arm at the end of the spindle – this is often solid, so you might have to just hacksaw through it.

3 Pull the arm off the spindle.

4 Disconnect the arm from the metal hoop attached to the siphon.

5 Undo the nut inside the cistern that holds the handle in place.

6 Now pull the handle out.

7 To fit the replacement, put the handle spindle through the hole and tighten up the inside nut. Notice that the spindle itself is covered by a threaded outer spindle which is sometimes fixed to the handle when you buy it, and sometimes you need to put this on first.

TIP *Sadly many older cisterns have a very small hole for the handle that doesn't fit those used on modern toilets. If this is the case you may have to shop around, but you'll usually be able to order a replacement.*

8 Fit the metal hoop at the top of the siphon to the handle arm.

9 With the handle in the right position fit the arm to the end of the spindle and screw it into position. Now check it works.

REPLACING A TOILET SIPHON

The classic indicator of a broken toilet siphon is that one turn of the handle no longer works. Over time, neither does two or three, until in the end it ceases to flush at all.

Older siphons usually stop working when the plastic sheet within them (which they rely on to lift the water and start the flush) develops a crack. As such, when you operate the handle you just lift a bit of broken plastic, the water runs through all the cracks and not much else happens. To solve the problem we need to either replace the plastic or replace the siphon:

Tools and materials
- Flat-head and Pozi screwdrivers
- Adjustable spanner
- Kitchen knife
- WD40
- Pump pliers
- Plumbers' mait
- Dust sheet

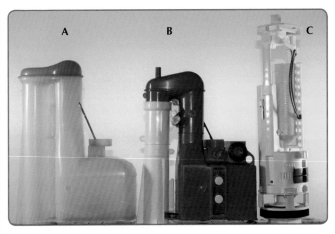

1 Lift the lid off the cistern carefully. If you have a toilet handle then the lid will just lift clear, but many push-button cistern lids are screwed into place and lifting the lid may break the mechanism.

TIP *If you are thinking of replacing the siphon because it's not flushing properly and, having lifted the lid, you have discovered a siphon that looks like C, then the siphon is unlikely to be the cause of your problems. Check the push buttons or, on some models, the tubes leading from the siphon to the buttons. If you still think it needs changing see 'Replacing Newer Siphons' in the next section.*

2 Have a look at what siphon you have. The odds are that you have type A, but just check again to see if you have type B.

The Dudley siphon can be taken apart by pushing out the yellow plastic screw near the top and then lifting the siphon head out.

At this point you just replace the broken plastic sheet. Sadly,

doing this with type A is an altogether more complicated process. Other manufacturers also make siphons that can be dismantled so check carefully at the outset, as it will save you a huge amount of time and effort.

3 Turn off the water to the toilet cistern and empty the cistern. You might be able to do this by just turning the handle several times, but you might have to resort to a sponge – you can remove the float to give yourself more room.

4 Disconnect the water supply by turning the tap connector at the base of the float valve anticlockwise. There are two types of float valve – side-entry and base-entry, in which the tap connector is found at the side of the cistern or just underneath it.

5 When you remove the tap connector check the state of the washer. If it looks a bit shabby, replace it now.

6 Disconnect the overflow pipe. This is most often at the base of the cistern on the opposite side from the float valve, but can be at the side. It has a plastic nut that can usually be turned by hand.

TIP *Toilets with old-fashioned siphons always need an overflow pipe. Newer toilets often don't, the siphon acting as an overflow instead. If something goes amiss in this instance the water will start to pour down the siphon and flow into the toilet pan.*

---

7 Disconnect the cistern from the wall. It's normally held on by two screws, so just undo these. Occasionally it'll be held in place by dollops of sealant. If this is the case you'll need to cut through it with a knife.

8 The cistern is held to the toilet by two butterfly nuts underneath the toilet. You can usually undo these by hand but if they've rusted into place you might have to resort to some WD40 and a pair of pump pliers.

9 The cistern should now lift off the toilet. Once off, lay it on an old towel – be prepared for a little water and rust to be released at this stage.

10 Here you can see the base of the toilet siphon, together with a rubber seal (called a doughnut seal for obvious reasons) and a metal plate with the two bolts that we've just undone loosely connected to it. More modern cisterns will often be held in place by bolts that pass through the base of the cistern, so you'll just see two bolt tails and a doughnut seal.

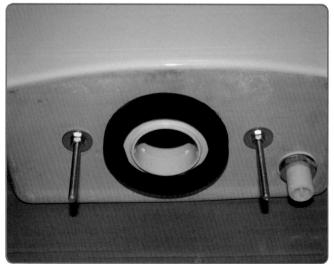

11 Remove the doughnut seal and set aside. Now use a large set of pump pliers to turn the nut at the base of the siphon anticlockwise.

12 Once you have this off remove the metal plate. This is often rusty, so put a dust sheet over the floor first.

13 Finally you can remove the toilet siphon by disconnecting the metal hook from the toilet handle and lifting the siphon out of the cistern. To effect a repair

you can either replace the broken plastic sheet or replace the siphon with one that doesn't need 13 steps just to remove the bloody thing, ie a Dudley Turbo siphon. To replace the siphon jump to step 15.

Remove the spring and the clamp at the base of the spindle. You can now replace the plastic sheet either with a new sheet designed for this purpose or with any bit of plastic that's been cut to the right shape.

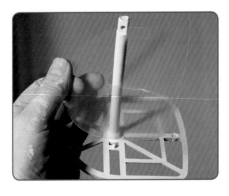

Now reverse this process to get the spindle back into the siphon.

15 You now need to put everything back together again. The siphon uses a rubber washer to create a watertight seal with the cistern. If you're refitting the old siphon check the state of this washer and replace it if need be. Also, use tissue or a cloth to clean the base of the cistern of debris that might affect the washer.

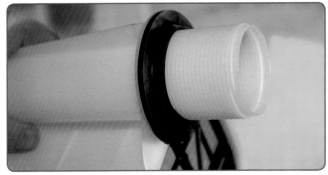

TIP *Whilst the height of most modern siphons can be adjusted, old siphons come in two set sizes. As such it's often best to take the old one with you when you buy a replacement.*

14 If you have opted to replace just the plastic sheet then you'll need to first remove the metal hook that connected the siphon to the toilet handle. With this gone you can slide off the two washers and then push the spindle down until the entire mechanism emerges out of the base.

TIP *Plumbers' mait is a type of putty that doesn't go hard. As such it can be used to make a washer if you can't find a proper one. To use, just dampen your fingers to stop it sticking to you, pull a bit out of the tub, roll it into a sausage and apply it round the item to be sealed.*

16 Connect the metal hoop at the top of the siphon to the toilet handle and check the handle is resting in the right position. Once you're sure push the base of the siphon through the hole in the cistern.

17 Put the metal plate back on. If your old one is all rusty then pop down to your local plumbers' merchants or DIY store and buy a new plate and doughnut seal pack. Check you have the plate the right way around so that you can put the bolts on it later.

18 With the metal plate in position, hold the siphon in place and tighten the siphon nut using a pair of pump pliers. Remember, we're just compressing some rubber to create a seal so there's no need to overdo it – if you do you'll probably end up cracking the plastic nut. Some siphons come complete with a spanner for this nut. If you're lucky enough to have one, feel free to use it.

19 Fit the bolts back on to the plate and then put the doughnut seal over the top of everything. It's always better to buy a new seal than risk using the old one.

TIP *Some toilets have a thinner, wider doughnut seal. If this is the case fit the seal to the toilet itself.*

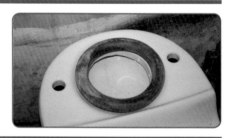

20 Gently lower the cistern back on to the toilet, making sure that the bolts are aligned with the holes in the toilet and that they don't fall off the plate as you lower the cistern into place.

21 Put the overflow pipe back into position and tighten in place – hand-tight is usually sufficient.

22 Tighten the butterfly nuts underneath the cistern, making sure that you have a rubber washer between the metal washer and the ceramic. Again, you're just pulling the cistern down and holding it in place so there's no need to over-tighten the nuts.

Tighten one nut a little, then tighten the other nut, then go back to the first again, *ie* try to draw the cistern down evenly.

23 Double-check that the washer's in place and then connect the tap connector to the base of the float valve. Often this is a case of a brass nut on a plastic thread so be very careful not to thread the plastic. Do it up by hand and then use an adjustable spanner to give it a final three-quarter turn.

TIP
It's often easier to only loosely fit the butterfly nuts in step 22 so that you can move the cistern a bit in order to get a good connection to the tap connector. Once the tap connector is on, tighten the butterfly nuts fully.

---

24 Turn the water back on and check for leaks once the cistern is full. Then check again when you flush the siphon. If all is dry, screw the cistern back against the wall.

It can take some time for water to leak from the base of the siphon, so check it again after an hour or so. If water leaks out when the toilet isn't being flushed the leak will be coming from the washer at the base of the siphon. If the leak only occurs when you flush the toilet the leak is coming through the doughnut seal, so take the cistern back off and check this seal is in the right place and that you've screwed the cistern down properly.

REPLACING NEWER SIPHONS
If like me, you think 24 steps is about 20 too many for replacing something as basic as a siphon then you might want to upgrade your siphon to one that can be dismantled, as discussed earlier.

Many modern siphons tend to work by just raising and lowering a rubber seal. It's rare for these to just stop flushing, but if they do it's usually because someone has ripped the cistern lid off

and taken half the siphon mechanism with them. In this case try to identify the remains of your siphon and buy a replacement.

These siphons, almost without exception, come in two parts and the main part of the siphon can be removed from the cistern by either twisting them a quarter-turn anticlockwise and lifting them from the siphon base, or by pressing in clips on the side of the siphon.

If this works you don't have to take the entire cistern out to replace the siphon mechanism. Within the trade this is known as 'a good thing'.

The most common issue with modern siphons is that they constantly leak water into the toilet pan. By a lucky chance we're going to look at this next.

My toilet is overflowing
Toilets overflow for two main reasons: either the ball valve within the toilet is no longer closing completely, so the water level keeps rising until it reaches the overflow point; or the siphon within the toilet is no longer sealing correctly and is letting water constantly leak out through the flush and into the toilet pan.

If you have an older style of toilet with a traditional siphon the odds are that the problem is the ball valve, and the water will be coming out of a pipe in the wall and dripping down outside the house.

If it sounds as if the toilet is constantly filling and water is running down from the rim of the toilet and filling the pan then you probably have a newer toilet and either the ball valve or the siphon isn't closing. It's not easy to see this effect without putting your finger – or ideally someone else's finger – on to the toilet pan wall and seeing the water build up around it.

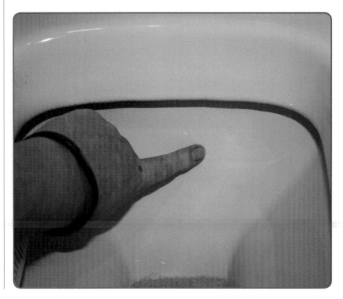

BROKEN BALL VALVE

Whilst ball valves come in all shapes and sizes they all work on the same principle – water comes into the cistern, which causes a float to rise. Eventually this float rises to such a height as to cause the valve to close.

In olden days the ball used to fill up with water and stop floating but this is so rare these days as to not be worth discussing. What causes most ball valves to fail is that the little rubber valve within them starts to split or crack.

With this in mind you could just replace this little bit of rubber, but it's better to check the state of the whole valve and, if it's seen better days, replace the entire thing. It's also a lot easier to buy an entire ball valve than it is to buy just the rubber valve.

REPLACING THE RUBBER BALL VALVE WASHER

Tools and materials
■ Small flat-headed screwdriver
■ Pump pliers
■ Replacement ball valve washer

1 Turn off the water to the toilet and test this by flushing the toilet and checking that it doesn't refill.

2 Every valve differs but usually they come with a nut at the end to which the float arm is attached via a little button. Undo this nut. Usually you can do this by hand but use a pair of pump pliers if need be.

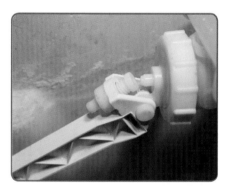

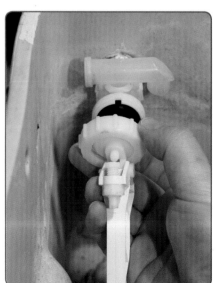

3 Carefully pull away the arm, getting ready to catch the button as it drops out.

4 You should now be able to

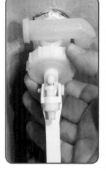

see the valve. Usually it's made of black rubber and sometimes it'll have a bit of white plastic built into it. Take note of just how it's arranged, as it will need to go back exactly the

same way. Take a small flat-headed screwdriver and gently prise the rubber valve out of its seating.

TIP *In this modern age, the easiest way to make sure you put something back together correctly is to use your mobile phone to take a photo of the arrangement before you touch anything.*

---

5 With that done, note which side is which on this bit of rubber, *ie* which face was pointing out at you when you removed it. Also be aware that there's often a top and bottom to these washers.

6 If you give the rubber a bit of a pull you should see what the problem is – usually there's a tiny split. If you can't see anything wrong you might want to consider replacing the entire ball valve.

7 Now for the hard part. You need to find an identical replacement for this rubber valve. This isn't as easy as you might expect, but once you've achieved it just put the valve back exactly as you found it and reverse the steps that got you here in the first place.

REPLACING THE ENTIRE BALL VALVE

If you have an old-fashioned ball valve, *ie* it has a long arm with a great big ball at the end, you might want to consider updating it for a quieter model. Either way the first thing you need to do is to get the old valve out.

TIP *New ball valves generally come in two sizes and the height given is from the base of the cistern to the top of the ball valve, so measure this height before you start and aim to buy a replacement that's as close to this height as possible. This isn't always feasible so some new designs have an adjustable height, which makes life a lot easier.*

Tools and materials
- Adjustable spanner
- Bowl
- Sponge
- 15mm tap washer

1 Turn off the water to the toilet, flush the toilet and check it isn't refilling. If your float valve connects to the side of the cistern jump to step 2, otherwise grab a sponge and get the rest of the water out of the cistern – unscrewing the ball at the end of the valve will give you more room.

2 The ball valve is connected to the water supply by a tap connector. You need to take an adjustable spanner and turn the nut on the top of the connector anticlockwise to loosen it. Once it's undone pull it away from the base of the ball valve. Check the state of the washer in the tap connector and replace it if it looks worn.

3 The valve is held in place by a nut on the outside of

the cistern. Use an adjustable spanner to turn this nut anticlockwise. If the ball valve is connected to the base of the cistern you'll probably get a bit damp at this stage so have a bowl or sponge ready.

4 The old ball valve can now be removed.

5 Every ball valve is slightly different so always reads the manufacturer's instructions first. However most, if not all, follow these basic steps....

The waterproof seal is made by a rubber washer squashing up against the ceramic of the cistern. Consequently this area of the cistern needs to be free of debris, so give the area a good clean.

6 Put the rubber washer on to the base of the ball valve and put the new ball valve in place, arranging it so that the mechanism is free to move up and down.

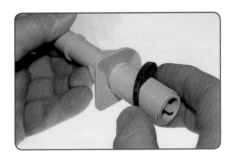

TIP *Often the washer supplied with a new ball valve is cone-shaped. If this is the case the pointy end of the cone points towards the ceramic hole.*

7 With the base of the new valve pushed through the hole in the cistern, put the nut on the base but don't tighten it yet. Whilst you're there, note that this nut often has a circular ridge that's designed to just fit into the hole in the cistern, ensuring that the ball valve is centred perfectly in the hole. To use this you need to make sure the ridge is facing into the cistern.

8 With the main nut only loosely on, hand-tighten the tap connector to the base of the ball valve. This is often a case of connecting a brass nut to a plastic thread, so there's a huge risk of threading the plastic. To minimise this, align both ball valve and tap connector (this is why we haven't tightened the nut on the base of the ball valve yet), push the tap connector hard up against the base of the valve and only use your fingers to turn the nut clockwise. If everything is aligned correctly you should easily be able to turn the nut on to the thread. If you encounter a fair amount of resistance you're probably not aligned properly, so undo the nut and start again.

Of course, you can make life much easier by only buying ball valves with brass threads on the base. They might be a little more expensive but they'll endure the trauma of being fitted and serviced far, far better.

9 Now hold the ball valve in place with one hand and tighten up the nut at the base of the ball valve with the other, using an adjustable spanner to give it a final three-quarter turn. You might want to rope in a friend at this point.

10 Still keeping the ball valve in place with your hand, use the adjustable spanner to give the tap connector a final three-quarter turn.

11 You now need to adjust the ball valve so that it closes at the right water level. Each type of ball valve has its own way of doing this. If you're fitting a traditional ball valve with a large plastic ball at the end then you adjust the height of this ball on the arm by loosening and tightening a screw. Newer ball valves are usually adjusted by turning a plastic screw – see 'Setting the cistern water level' earlier in this chapter.

12 Now turn the water back on and check for leaks. If all is dry replace the cistern lid.

The siphon is leaking

Newer siphons work by lifting and dropping a large rubber seal over the hole in the base of the cistern. They often use the weight of the water in the cistern to push down on this rubber seal and make it watertight, and as such they'll often leak slightly whilst the cistern is filling but become totally waterproof once full. These types of siphon fail either because the rubber seal wears away or develops bubbles in the rubber, or, as is more likely, because debris in the cistern breaks the seal.

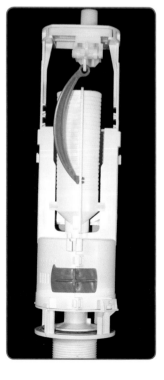

More often than not this debris is introduced into the cistern by the householder in the form of those little blocks that turn the flush water interesting colours. If you have a modern push-button toilet

don't put these blocks into the cistern – use the under-the-rim variety if you can't face life without coloured flush water.

If you think the siphon might be the cause of your toilet overflowing then:

1 Turn off the water to the toilet.

2 Flush the toilet.

3 Remove the top of the siphon by either holding it close to its base, giving it a quarter turn to the left and lifting it out, or looking to see if there are any clips on the side of the siphon (if your siphon is green and white it may well have clips). If there are press these in at the top and lift the siphon out.

Note that each design differs so if the guidance given above doesn't work read the manual that came with the siphon (assuming you ever had one), or try to find some information about it online before attempting to remove it.

4 Turn the siphon over and have a look at the rubber seal on the base. It's not unknown for these to develop bubbles that stop it forming a watertight seal. If you think it's damaged take it to your local plumbing merchants and order a replacement – take the whole siphon so they can identify the manufacturer and model for you. It's unlikely that anyone will hold a replacement seal in stock so don't throw the old one away quite yet.

If the seal looks fine check for debris in the base of the siphon that's still in the cistern.

 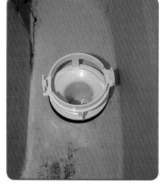

5 Clean out the cistern with a sponge to remove all the debris.

TIP *If the washer looks damaged you can often just take it off, turn it upside down and refit it. This mightn't work for long but it will give you enough time to order a new washer.*

6 Fit the top of the siphon back on to the base, giving it a twist or pressing in clips as appropriate.

7 Turn the water back on and check that the leak has stopped. If you couldn't replace the seal straight away and the toilet is leaking a lot of water, turn off the toilet at the isolation valve. If your toilet doesn't have one, then fit one – see 'Isolation valves' in Chapter 7 and basic techniques in Chapter 6.

The tanks in the loft are overflowing

If you have tanks in your loft they'll be connected to one or more overflow pipes that terminate just under the guttering of the house and are designed to allow water to drip or pour down the outside wall in a manner that will engage your eye and warn you that something in your loft has gone wrong and needs to be sorted asap.

There are often two tanks in the loft; the larger is for your hot water and the smaller is for your central-heating system. By looking at the location of the overflow pipe on these tanks you should be able to deduce which one has an issue.

Usually the problem is caused by the ball valve in the tank failing to close properly and you can replace the washer or the entire ball valve in much the same way as we discussed earlier for the toilet ball valve. Usually the ball valves are in a heck of a state by the time they fail, so even if it's only the rubber washer that's failed it's usually better to replace the entire unit. (See 'Replacing the rubber ball valve washer' earlier in this chapter.)

Tools and materials
- Adjustable spanner
- 15mm fibre washer

1 Turn off the water to the tank.

TIP *If you have an old galvanised tank be very careful! Before you do anything check out the condition of the tank and if you think it looks corroded consider replacing it. Once they start to corrode they can just fall apart at the slightest touch, so don't do anything else until the tank's been replaced.*

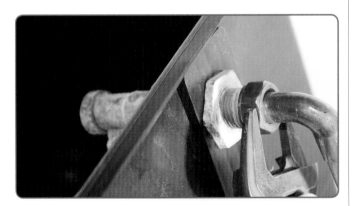

2 Disconnect the tap connector by turning the nut anticlockwise.

3 Remove all remnants of the old washer and fit a replacement.

4 Undo the nut holding the ball valve in place and remove the old valve. There's often a metal or plastic plate attached to the base of the ball valve. This is to stop the tank flexing when the ball valve moves and isn't usually fitted these days. Having said that, they do no harm, so if you have one reuse it.

5 Put the replacement in position. Note that there are two nuts – one goes on the inside of the tank and allows you to adjust the distance between the valve head and the side of the tank, the other goes on the outside of the tank and holds the valve in place. Adjust the first nut by hand until the distance of the valve head is roughly the same as the old one and then put the valve in place so that the valve arm drops vertically and not at an angle.

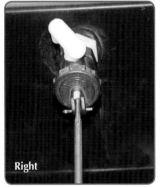
Right

Wrong

Now hand-tighten the outside nut before giving it an extra three-quarter turn with an adjustable spanner.

Note that there are no rubber washers

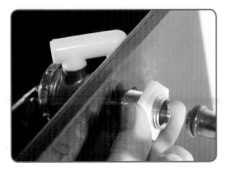

involved to create a waterproof seal. This is because the water level should never get this high – it will pour out of the overflow pipe first.

TIP *The replacement valve should always have a long arm on it – ideally a metal arm – with a plastic ball at the end, ie don't use the smaller-armed toilet valves in your loft tanks. If you fancy getting technical you should pop down to your plumbing merchants and ask for a 'BS1212 Part 2 ball valve'.*

6 Check that the new washer is still in place on the tap connector and that there are no older washers underneath it. Now hand-tighten the nut to the threaded base of the ball valve. Once tight give it an extra three-quarter turn with an adjustable spanner.

7 Turn the water back on. Check that nothing is leaking and check that this cures the overflow problem. If it doesn't, read the warning on this page.

⚠ **WARNING**

If the ball valves have been replaced and the tanks are still overflowing then you have a problem with either your central heating or your hot-water cylinder, and the overflow is caused by hot water being discharged into these tanks. In both cases you're better off getting a professional in to check things out.

My taps are dripping

In the good old days this was just a matter of tightening a gland nut or changing a washer. Alas, taps vary enormously these days and you can never be too sure what you'll find when you take them apart. Fortunately, there are a few clues:

- If your tap operates by lifting the handle to turn the water on and then twisting it to adjust the temperature you have a monobloc tap with a ceramic quarter-turn mechanism.
- If your tap has two heads that turn on and off by turning them just a quarter turn, then – not surprisingly – you have quarter-turn taps that also use a ceramic mechanism.
- If you have to turn the tap heads round and round to turn them on and off then your taps undoubtedly use rubber washers.
- If, whilst turning the taps round and round, you notice that the tap head rises or lowers then you have a rising spindle tap. These have a gland nut which is another common cause of leaks – although the water will swell up around the tap itself and not be dripping out of the spout – see 'Repacking the gland nut' (Chapter 7).

CHANGING A TAP WASHER

The first problem you have is getting at the washer, which can vary from tap to tap. What you're aiming to reveal is a large nut that allows you to unscrew the top of the tap from the main body. As this screw is deemed unsightly it's invariably hidden under either the tap head itself or a little plastic or metal cap.

Tools and materials
- Adjustable spanner
- Flat-head and Pozi screwdrivers
- Cloth
- Pump pliers
- Box spanner

1 Turn off the hot or cold water, depending on which tap has the problem. If you have any doubts, *eg* you have a single mixer tap, turn off all the water. Having done that, test it by opening the tap and waiting until the water stops flowing – if you have a monobloc tap make sure both hot and cold are opened by lifting the handle and setting it to the middle.

2 At the very top of the tap head there's usually a plastic cover. This can be lifted off either by using a fingernail or a small flat-headed screwdriver. With newer taps you may find that the head is removed via a grub screw, which is often hidden behind a plastic plug.

3 This should reveal a screw head, which you undo.

4 You can now lift the head off the tap.

5 This should reveal the nut. However, it may still be covered by a plastic or metal shield. If so this usually twists off by turning it anticlockwise.

Note that this shield is often made from very soft metal and can easily be damaged, so use WD40 to make removal a bit easier and a cloth and a set of pump pliers as a last resort.

6 The main nut should now be revealed. Undoing this nut can be a real endeavour. The best tools for it are a box spanner and an adjustable spanner. Put a cloth over the tap spout.

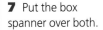

7 Put the box spanner over both.

8 Put the adjustable spanner on the nut and, using the box spanner as leverage to keep the tap in

place, turn the nut anticlockwise. If this doesn't work, spray it with a bit of WD40, leave it for five minutes and try again.

Still no luck? Then ask a friend to hold the box spanner and give it another try.

Still failing? Then try hitting the adjustable spanner with a hammer to give it a short sharp shock. If this still fails then I'd either just give up and look at changing the taps altogether (see Chapter 10) or maybe try heating the nut using a blowtorch or try a freeze-shock spray.

9 Having undone the nut the tap head should lift off to reveal the washer, which is at the end of a spindle that rises and falls as the tap head is turned.

TIP *As the washer wears down people have a habit of turning the tap tighter and tighter, rather than just replacing the washer. This can often cause the spindle to fall out of the tap mechanism altogether and as a result the tap neither opens nor closes properly.*

If this is the case you'll notice that the part holding the washer falls out as you remove the tap body, or even remains in the tap. To deal with this just follow steps 1 to 9, then add a little dab of silicon grease to the spindle (make sure it's only a dab, as you'll just clog up the mechanism if you use too much), push the bit containing the washer back into place and turn the tap head anticlockwise to draw it back into the tap body.

10 Remove the old washer – if there's still one there. To do this you can usually just pull the washer over a little brass mushroom-looking affair, but occasionally there's a small screw holding the washer in place, which needs to be undone.

11 If the washer isn't in too bad a state you can just turn it over and reuse it. However, having gone to all this bother to reveal it you might as well replace it entirely by either pushing the new washer over the mushroom or putting it in place and tightening up the little screw again.

When selecting a new washer you should obviously try for an exact match. Sadly this isn't always possible, in which case always opt for a washer that's slightly smaller than the original. If the new washer is too large it tends to get stuck in the tap body and eventually falls off, causing all sorts of problems. If you can't find a suitable washer you could use a utility knife to pare down a larger washer, but this should be a last resort.

12 Check that the inside of the tap body is free of debris – have a look into it if you can or put your finger in and feel around; often there are bits of the old washer stuck here which need to be removed.

13 To refit the tap head just follow steps 1–8 in reverse. Once all is back in place turn the water on and check that the new washer works and that you're free from leaks.

REPAIRING A QUARTER-TURN CERAMIC TAP

Most modern taps have turned their backs on rubber washers and opted for a quarter-turn ceramic mechanism. In this mechanism there are two ceramic discs that turn on top of each other. Turn one way and they block the tap completely, turn them a quarter turn the other way and a hole in each lines up and water can flow through.

The idea was that these would never leak because the ceramic was too hard to be damaged. Sadly this has turned out to be nonsense; in hard-water areas limescale can get between the two discs and, over time, score the surface. Once scored the mechanism tends to gently drip when closed.

Another cause of quarter-turn taps leaking is people turning them far too hard – you should be able to open and close these taps using just one little finger. Constant abuse causes the ceramic discs to turn

slightly past one another when the tap is closed, so that a tiny hole appears and water drips out of the tap until you turn the head back a tiny fraction.

To repair a ceramic quarter-turn tap follow steps 1 to 9 in 'Changing a tap washer'. This should reveal the ceramic tap cartridge.

Note that the base of this cartridge has a rubber washer. This is there to seal the mechanism against the base of the tap. It can be changed, but since this washer never turns it rarely if ever gets damaged.

Nope, your best bet here is to just change the entire cartridge. Sadly, there's not a universal standard for these so you'll have to pop down to a plumbing merchants and show them the old one. Beware that the vast majority of the cartridges will appear identical until you try to fit the old tap heads, at which point you discover that the head of the cartridge doesn't fit the head of your tap. With this in mind take the tap heads with you when you visit the merchants. You might be lucky enough to own a tap that has a manufacturer and possibly even a model name printed on it. If this is the case give the manufacturer a ring, or visit their website, and order some spare ceramic cartridges.

NOTE *The washer at the base of the ceramic cartridge is invariably coloured red or blue – red for hot, blue for cold. You need to make sure you use the right replacement otherwise you'll have to turn your taps the wrong way to close them, which just makes life horribly complicated for all concerned.*

Of course, the hot and cold cartridges don't fail at the same rate – the hot cartridge is the one that usually fails because more limescale is deposited in hot water than in cold. It should therefore come as no surprise to discover that you have to buy a set of cartridges every time just one fails.

Once you have a replacement follow steps 9 to 1 in 'Changing a tap washer' to put it all back together again.

If you have lever taps you need to make sure you have the lever heads aligned properly. To do this you're best off turning the water back on before screwing down the tap head. This way you can put the tap in the closed position, align the head in an aesthetically pleasing manner then check them when they're open and closed before screwing the head down and locking it in place.

REPAIRING A SINGLE LEVER MONOBLOC TAP

This is a single tap that delivers both hot and cold water, and the only real difference here is the design of the ceramic cartridge – which varies enormously from one manufacturer to another. With this in mind you first need to identify your tap and order a replacement cartridge. Once this arrives, just follow steps 1–9 in 'Changing a tap washer', although the cartridge ought to arrive with a set of installation instructions anyway.

TAP RESEATING TOOL

In step 12 of 'Changing a washer' I mentioned checking the inside of the tap. A tap washer can only work if the hole that it closes over has smooth edges, but over time this smooth edge can get pitted. If this happens the tap will leak regardless of the condition of the washer.

The only time I've ever personally come across a tap where this had happened the tap itself was in such a bad shape as to need replacing anyway. I can only assume that modern manufacturing techniques mean that the brass is very unlikely to pit these days and that this scenario is now very rare. That said, everyone and his mate seems to sell this little tool, so I can only assume that there's still a huge demand for it.

So if you think that pitting is the cause of your leak, the aptly named 'tap reseating tool', is the tool to resolve it for you. It can be purchased in most plumbing merchants and DIY stores, and to use it you first need to follow steps 1–9 in 'Changing a washer'.

1 You now need to select the correct grinder and bush to fit your tap – the tool shown has a series of different threads built into it so that you don't have to fit a specific bush. The grinder will be the one that only just fits inside the tap and the bush will be the only one that fits the thread at the top of the tap.

2 Put the tool into your tap and tighten it in place.

3 Now turn the handle clockwise whilst pushing down. The grinder will start to scrape away at the seating, removing the blemishes that have been causing the leaks.

4 After every half-dozen turns undo the bush and take the tool out of the tap. Check the state of the seating to see how you're doing – use a torch to see the state and use your finger to feel for any blemishes. If you think all looks good then put the tool back in place and give it a few light turns just to polish up the seating. If you have a vacuum cleaner handy it pays to suck out any swarf debris as this stage.

5 Now follow steps 9–1 in 'Changing a tap washer'; but before you do, make sure the tap is in the open position – you don't want to ram the new washer down on to the sharp brass flakes you've just littered across the inside of the tap.

6 Turn the water on and let the flow wash out the inside of the tap and flush all the debris out before you turn the tap off.

TAP RESTORER KITS

Rather than buy a complete set of new taps it's often cheaper to replace just those bits that have worn out. Tap restorer kits contain a new tap mechanism and new tap heads.

One big advantage of these kits is that you can use them to convert ordinary taps into quarter-turn lever taps. These have levers on top that make them much easier to operate. To fit them just follow the steps in 'Changing a tap washer'.

LEAKY SPOUT

Most modern kitchen taps have a long swan-neck spout that can be moved from side to side. Sadly these have a habit of leaking at the base of the spout, causing limescale to build up around them. This can be repaired as follows.

Tools and materials

- Small flat-headed screwdriver or Pozi screwdriver or set of Allen keys
- Silicon grease
- O-rings

1 Turning the taps off will isolate the water whilst you make this repair, but you might want to play safe and turn off your hot and cold water anyway.

2 There should be a small grub screw at the back of the tap, close to where the spout meets the main tap body. This can be a flat-head, Pozi or Allen key fitting, so unscrew using the appropriate tool. Occasionally there isn't a screw and you have to turn the tap spout to a certain angle instead, usually 90° to the base, and then pull it out.

3 With the grub screw removed pull the spout out of its seating. You should now be able to see a number of rubber O-rings at the base of the spout. Usually one or more has been worn down or cracked. You can buy replacement O-rings from most plumbing and DIY stores, or if you know the name of the manufacturer you can order them online – always the best option, as you can guarantee they'll fit properly.

4 Remove the old O-rings, using a small flat-headed screwdriver to lift them from their seating.

5 Rub the new O-ring in a small amount of silicon grease to lubricate it.

> ⚠ **WARNING**
>
> Don't use petroleum-based greases such as WD40 or Vaseline on O-rings, as these tend to corrode rubber. This is an especially important tip if you don't wish to enlarge your family.

6 Fit the new O-ring in place.

7 Push the spout back into place. Make sure it's fully in before tightening the grub screw. Check that the spout still swivels OK and if everything looks well open the tap and check that all is now dry.

Dealing with an airlock

If you drain down your hot-water system there's a chance that you'll get an airlock. This is caused when a bubble blocks the pipework and the water pressure is insufficient to push it out of the way. As such it generally only happens with gravity-fed hot-water systems, and it depends on how your hot water has been plumbed as to whether this will be a rare or commonplace occurrence.

If you do get an airlock there are a number of things you can try to dislodge it. The easiest way is to use a water vacuum pump.

Option 1
With only the one tap open, put the vacuum on suck and apply the hose to the end of the tap that's blocked. If this doesn't work open the other hot-water taps and try again.

Option 2
Still no joy? Then apply the suction to the vent pipe over the top of the cold-water storage tank – this is described in Chapter 2.

Option 3

If all this has failed try firing mains cold water into the tap that isn't working by connecting a hose to the cold water (usually the outside tap) and pushing the hose into the end of the tap – expect to get a little damp when you're doing this. Again, if it doesn't work try forcing cold water into some of the other hot-water taps, and finally the vent.

Option 4

If you still have an airlock go back to the hot-water stop tap (see Chapter 2). The airlock is often just above this tap and the easiest, but far from relaxing, way of dealing with this is to:

1 Ensure that the hot-water stop tap is open.

2 Put a bucket under it.

3 Open up the nut on the top of the tap. For reasons that will become clear, this should be regarded as a last-ditch solution.

As you undo the top nut you'll get a steady trickle of water out of the pipe and hear a number of ominous glugs. If you're lucky one of these glugs will be the airlock finally shifting, in which case you'll hear the whoosh of running water. When this happens, quickly tighten up the nut again.

If the water still isn't flowing properly you might need to actually pull the pipe out, or almost out, of the fitting. At this stage there's usually a rapid series of glugs and the water suddenly starts racing down the pipe towards you. If you're quick enough you can get the pipe back into the fitting and tightened with only the smallest of leaks, but you need to be fast ... and lucky. Worst case scenario is you get a little wet, fill the bucket and might have to mop up a bit afterwards, but you ought to have finally cleared the airlock!

To be honest, when things get this desperate you might want to call in a professional. If nothing else it provides peace of mind, and you'll have someone else to blame if it all goes pear-shaped!

Fitting a new cold-water storage tank

Though this isn't something you'll do very often it's handy stuff to know. It's also very easy to do right and even easier to do wrong.

There are two main reasons why might you need to change your tank:

YOU HAVE A GALVANISED STORAGE TANK

These are made from steel covered with a coating of zinc to stop them rusting. They stopped being used for cold water storage shortly after the Ark was launched, but many older homes still have them. If you have one, you really ought to change it.

The problem with these tanks is that the zinc eventually gives up the ghost and the steel underneath starts to rust. A tank can last for years in this precarious semi-rusted state, until one day you sneeze within two feet of it and the entire thing collapses and floods your home.

YOU WANT TO FIT A MIXER SHOWER

The standard cold-water storage tank is about 114 litres (25 gallons), which is usually large enough to supply all the hot water in your home. However, a mixer shower requires cold water from this tank as well, and 114 litres often isn't enough to cope with the demand. The easiest way around this is to connect a second tank to the first, but you might decide that the old tank has seen better days and opt to replace it.

See Chapter 11 for more on installing a mixer shower.

Tools and materials
- ■ Tape measure
- ■ Abrasive strip
- ■ Water-resistant timber (to support tank)
- ■ Drill
- ■ Hole-cutter
- ■ Ball valve
- ■ Isolation valve
- ■ Tap connector

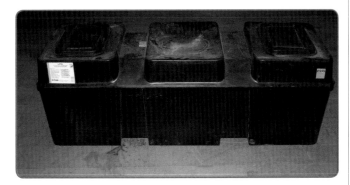

1 Measure the size of your loft hatch and buy a tank that's going to fit through it. For smaller lofts they do what's called 'coffin' tanks that are easier to fit through small access hatches.

When you buy the tank make sure you also by the appropriate Byelaw 30 kit, also known as the Reg 16 kit.

TIP *If the hatch is really small then buy a few small tanks and link them together. Use 22mm push-fit tank connectors set as close to the base of the tanks as possible and connect using 22mm pipework. Try to keep the tanks as close together as possible.*

2 Prepare the tank before you go up into the loft. First check the diameter of the tail of your float valve and use a hole-cutter to drill a hole just wide enough to accept it. Drill this hole about 40–60mm from the top of the tank.

3 Fit your new float valve into this hole. Use the front nut to adjust the distance from the back of the valve to the tank side and use the outside nut to tighten the ball valve into place.

Make sure the arm of the ball valve drops down vertically and isn't at an angle. Note there are no washers attached to the ball valve – this is because the water level should never reach this point. Remember, don't use short-armed float valves in your loft water tanks!

4 In your Byelaw 30 kit there's an overflow pipe. Check the width of the tail for this pipe and drill another hole, just wide enough to accept it, at least 20mm lower down from where you've fitted the ball valve.

5 Fit the overflow pipe. Note that it has an extension pipe that, once the tank is full, dips down into the water. This stops freezing cold air blowing into the tank during the winter – see Chapter 5 for more details.

6 In an ideal world at least one of the outlet pipes should be at the bottom of the tank. This is so that sediment, which could harbour bacteria, doesn't build up in the tank. Alas, this is often difficult to achieve, but at least aim to put the outlets as low as possible.

At least one outlet should also be on the opposite side of the tank from the ball valve. This is so that water flows across the tank and you don't get any stagnant pockets of water. If you're fitting multiple tanks together always ensure one outlet is taken from the tank farthest from the ball valve.

TIP *Check the inside of the tank before you drill any holes, as strengthening structures are often built into it. If you drill through one of these you won't be able to fit a connector and will have to go out and buy a new tank.*

Use 22mm push-fit tank connectors as these are much easier to fit, and if you're fitting hot- and cold-water outlets make sure the cold water one is lower than the hot. This is so that if you run out of water, you run out of hot first. Do it the other way around and you risk scalding yourself if the tank ever empties.

Measure the diameter of the tank connector tail and drill a hole just big enough to accept it.

7 Clean the edge of the hole with an abrasive strip.

8 Push the tank connector through the hole and tighten it by hand.

9 That's all the holes you need to drill, so turn the tank upside down and clean out all the little bits of plastic swarf – these have a habit of getting into the outlet pipes and blocking them, so be thorough.

10 Now go up into the loft and build a platform to support the tank, bearing the following in mind:

- The higher up the platform is, the more pressure you introduce to the system.
- Never build the platform out of ordinary chipboard or MDF, as these blister and fall apart when wet – not ideal characteristics for something supporting a large container of water. Use floorboards or 19mm plywood or chipboard that's been treated and can cope with wet conditions.
- Ensure that all of the tank's base is supported – don't just lay it over bare joists.

The tank will weigh about 100kg (over 15 stone) once full, so make sure you're happy to jump up and down on the platform, possibly with a friend, before you put the tank on it.

> ⚠ **WARNING**
>
> When cold, plastic tanks are very strong. However, the cold-water storage tank also acts as an overflow for your hot-water cylinder, and if something goes wrong with the cylinder hot water starts to pour into the tank. As the water gets hotter the plastic becomes softer. If it's fully supported this shouldn't pose a problem, but if it's just resting on joists the tank might collapse and drop 114 litres of boiling hot water into the room below. This has happened, and children have been killed as a result.

11 With the tank on its platform you can now connect the ball valve to the water supply. First turn the cold water off and drain the system down by opening the kitchen tap. You might have to adjust the cold-water pipework to make the connection. If this is the case see Chapter 6, where basic techniques are discussed.

Fit an isolation valve to the cold-water supply close to the tank. Then fit a tap connector to the base of the ball valve, making sure there's a washer in place. Alternatively fit a tap connector that comes with an isolation valve already built into it.

12 The overflow from the tank needs to run downhill in 19mm plastic pipe and exit the house.

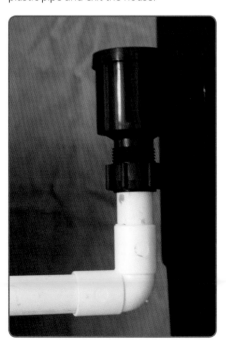

13 The hot-water vent pipe needs to rise up above the level of the water in the tank by about 450mm and then drop down into it – it shouldn't touch the water when the tank is full.

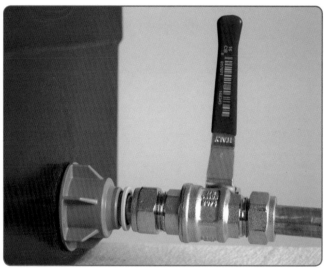

14 For the outlet pipes, you can – and should – fit a full-bore level valve on to the hot- and cold-water supplies. You need to be able to get at these valves in an emergency, so the hot is usually fitted just above the hot-water cylinder. Fit the cold one wherever seems best. Label them so that everyone knows what they do.

> ⚠️ **WARNING**
>
> If you're replacing the CH feed and expansion tank don't fit a valve to the outlet pipe. See chapter 12 for why this is.

15 You should now be ready to fill the tank, so turn the water on, open the isolation valve and let the tank fill. Once full check the water level – the base of the overflow pipe should be at least 25mm above the water level. Check everything for leaks and then follow the instructions on 'Insulating your storage tanks' in Chapter 5.

9 IN THE GARDEN

Fitting an outside tap   113

Garden irrigation system   116

Fitting a water butt   117

The most common plumbing item we find in the garden is an outside tap, and we'll be discussing how to fit these shortly; but before we get there it's important that you understand some of the rules about outdoor plumbing and why they're there. They're all common sense really, but as you might have discovered already, common sense is pretty thin on the ground most of the time.

Firstly, if you want to run pipework under your garden run it in plastic pipe, as copper will corrode over time and leak.

As much as is possible, run continuous lengths of pipework as it's the joints that usually leak, so the fewer of these the better. You can buy plastic pipe in long rolls and dispense with joints altogether, but whatever you do, don't use ordinary brass compression joints underground.

Make sure the pipework is buried between 750mm and 1.35m deep. There are a number of reasons for this:

■ Any shallower and you're likely to dig through the pipe whilst gardening.
■ At and below 750mm the ground doesn't freeze, so your water will remain nice and cool regardless of what the weather is doing on the surface.
■ Different utilities such as waste, electric and gas are buried at different depths, so keeping to the range given above ensures that your water pipework doesn't come into connect with any other utilities.
■ If you're forced to lay your pipes any shallower make sure the pipework is insulated from the cold. Also, make sure it's protected from damage by spades etc by putting it inside a larger diameter pipe or putting some sort of protection over it – a concrete slab, for example. It's also sensible to write on this protection something along the lines of 'Warning: water pipe below'.
■ As the pipework rises up to whatever outlet you're connecting it to remember to insulate it from the 750mm-deep trench all the way up to the outlet.
■ Lay the pipework in sand so that it doesn't get pierced by sharp stones etc and isn't damaged by any subsequent subsidence.
■ The water company goes to a lot of effort to ensure that your water is pure and drinkable. As such they take great exception to people sullying it with soil and other dangerous impurities and aren't averse to imposing huge fines. Consequently you need to ensure that all outdoor water supplies have valves attached to stop any dirty water being sucked back into the mains pipework. These are usually called 'check valves' and 'double-check valves' and are generically referred to as 'back-flow prevention devices'.
■ Finally, don't lay water pipes through, under or near septic tanks. In a sane world I wouldn't have to mention this, but sadly it isn't, so I do!

Fitting an outside tap

There are two things we need to be aware of when fitting outside taps. Firstly they're prone to freezing, and secondly they can suck dirty water back into the mains supply.

Some DIY stores sell little 'outside tap' kits that can apparently be fitted by chimps with little or no difficulty. Personally I don't like them because the water flow can be very restricted. That said, they're dead easy to fit, so give them a try if you're in a hurry. If you'd prefer to fit the tap 'properly', then select a location and follow these steps:

Tools and materials
■ Stop tap
■ Hose union backplate
■ 16–22mm drill bit long enough to go through a cavity wall
■ 6mm drill bit
■ Hammer drill
■ Rawlplugs
■ PTFE tape
■ Small piece of tape
■ Adjustable spanner and pump pliers
■ Marker pen
■ Level
■ 15mm pipe clips
■ Double-check valve
■ Isolation valve
■ Pipework and fittings of choice – copper or plastic

1 Buy a stop tap. If this is a new installation you need to buy one that doesn't have a check valve built into it, because these can fail if they freeze. If you're replacing a tap you can replace like-for-like, but it's always best to fit the check valve inside the house. The inbuilt check valve looks just like a small screw coming out of the base of the tap.

2 Drill a hole through the wall to accept a 15mm pipe.

3 You want to keep the external pipework to a minimum, so if possible use a 'hose union backplate'. This is a piece of copper long enough to go through most external walls that has a threaded connector at one end.

4 To connect the tap to the backplate, wrap PTFE around the thread of the tap. Wrap it clockwise, *ie* with the thread, and take it off the bottom of the roll, as this makes life much easier. Use at least four wraps but don't be daft about it.

5 Use an adjustable spanner to tighten the tap on to the backplate.

TIP *If you opt to use a backplate elbow you might find that having tightened the tap fully it's at an odd angle. Turning the tap back to get it at the right angle will usually mean it leaks. The only way around this is to undo it completely, whilst counting the amount of turns. Once undone, reapply the PTFE and tighten it up until you reach the last turn where you can still position the tap properly and stop there.*

6 Cover the open end of the copper tube with some tape to stop debris getting inside it, then thread it through your hole. Check that it's long enough, *ie* it reappears inside.

If your pipe isn't long enough use a backplate elbow and a longer length of pipe. What you don't want to do is attach two pipes together with a joint that will be hidden inside the wall. If you do this Sod's Law clearly states that the joint will leak and fill the cavity entirely unnoticed until rising damp starts appearing all over the place.

7 Turn the pipe it so that the tap is in the right position and use a marker pen to mark the position of the three holes.

8 Use a hammer drill to make three 6mm holes, fit rawlplugs and screw the backplate into position.

9 Go indoors and work out how you're going to connect this pipe to your cold-water supply, bearing in mind that you need to fit an isolation valve (so that you can turn the tap off in winter) and a double check valve (so you can't contaminate the drinking water). There's no hard and fast rule but it's always best to have the check valve closest to the tap with the isolation valve just behind it.

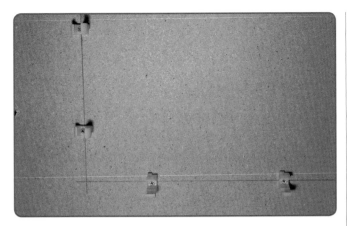

10 It's often easiest to mark the pipe run on the wall in pencil then fit pipe clips to the wall to support the pipework.

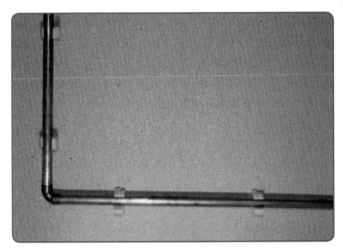

11 Put the pipes in place then mark and cut all the joints – see Chapter 6 regarding how to use copper and plastic pipework.

12 Fit the double-check valve, ensuring that the arrow points in the right direction, *ie* towards the tap.

13 Fit the isolation valve just behind the check valve, also ensuring that the arrow is pointing in the right direction.

14 Turn off your cold-water supply and drain it down. You might find that you still have some water in the pipe so put a bowl or a sponge underneath.

15 Cut into the pipework. Usually you'll want to fit a tee fitting at this point, in which case you'll need to cut out a

section of pipe about 15mm wide, although there's often enough movement in the pipework to create this gap by simply pulling gently on the pipework. Most of the time the connection to the cold water is made under the sink, and the cold pipe to the taps often has an isolation valve on it already. You need to make sure your tee is below this isolation valve, as you don't want the outside tap to be closed when the kitchen tap needs to be isolated.

16 Fit your joint into place. If you opted to use soldered copper bear in mind that it's unlikely to work if there's any water still in the pipework. If you can't get the water out opt for a compression or push-fit fitting.

17 Check the entire pipe run, making sure all your joints are complete, *ie* push-fit fittings have been checked to see if they're fully in and any locking mechanisms have been set, compression joints are tight and soldered joints have some solder showing. Having visually checked everything, make sure that the outside tap is shut. Only then should you turn the water back on.

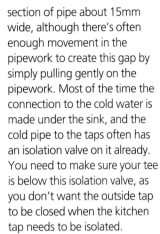

REPLACING AN OUTSIDE TAP

Many older outdoor taps come with a double-check valve built into the tap itself. The problem with these is that the check valve could fail in bad weather conditions, so they changed the rules and insisted that all new outside taps should have a separate check valve fitted indoors. BUT older taps were allowed to keep their built-in check valve, and if you replace one of these taps you can replace it like-for-like, *ie* with a tap with a built-in valve. However, if you opt to replace such a tap with one that doesn't have a built-in valve you MUST fit a double-check valve inside the house, as close as possible to the outdoor tap.

Tools and materials

- New stop tap, with or without a built-in check valve
- Adjustable spanner
- PTFE tape

1 Turn off the cold water – the outside tap should have an isolation valve fitted so you ought to be able to turn the water off at that point without affecting the rest of the house. Open the tap to ensure that the water is off.

If your existing tap doesn't have an isolation valve, fit one! This valve allows you to turn off the tap in winter and eliminate the risk of flooding due to frost damage.

2 Use an adjustable spanner to undo the old tap. As you undo it count the number of full rotations needed to get it out.

3 Take the new tap and wrap PTFE tape around the thread in a clockwise direction. Use at least four full wraps.

4 Put the new tap on to the backplate and tighten it, using the same number of turns as you needed to remove the old tap. (PTFE will keep a joint dry even if it isn't completely tight, but not if you've fully tightened it and then loosened it off. This is why we're counting the number of turns.)

5 With the tap in the right position, close the tap and turn the water back on. Check the back of the tap for leaks. If you think there's a leak repeat the steps but use more PTFE this time, or maybe even smear a little jointing compound over the thread – remember to use jointing compound suitable for drinking water.

Garden irrigation systems

Each one of these differs from the next, so we're not going to cover the installation procedure. What we do need to cover are the precautions you need to take to ensure you don't contaminate the water supply. Sometimes the manufacturers will supply and state what 'backflow-prevention' mechanisms are needed, sometimes they'll just say something along the lines of 'ensure that you conform to whatever water byelaws and regulations apply'. Most of the time you'll need to fit a double-check valve, but it's best to call or email your water company and actually ask them the question. If you don't fancy that then maybe ask a professional plumber to make the final connection to the water supply for you.

Fitting a water butt

If, like me, you're on a water meter your wallet probably itches every time you inundate your lawn with water from your outside tap. It's also a waste of water, and there's a simple way of avoiding this wastage: fit a water butt. These come in all shapes and sizes but the principles are the same for all of them:

- Pick a downpipe from your guttering.
- Cut into it and redirect the pipe into a water butt. This is usually achieved via a rain diverter supplied with the kit, but you can simply divert the entire downpipe into the water butt.

- Take an overflow pipe from the top of the water butt and reconnect this to the base of the downpipe.

If you are using the diverter supplied with your water butt you may find that there is no need to fit an overflow. Providing the diverter is fitted lower than the top of your water butt it will stop filling the butt as soon as it's full. With this in mind, always check the height stated in the installation instructions or, if like most installation instructions they're nigh on impossible to understand, put the water butt into position and then mark the height for your diverter.

Most water butts come with all the instructions you'll need, but here are a few things to think about before you buy one:

- The only tools you generally need are a tape measure, a hole-cutting set and a drill. Always double-check the size of hole you need before drilling it. The easiest way is to see what size bit the connector you're going to use will just fit into.
- If you have cast iron downpipes, making the connection to the water butt is going to be a nightmare. So much so that you might want to reconsider the idea altogether, or have your guttering updated, or bring in a professional.
- There's no standard shape and size of downpipe so you might have problems connecting to your pipework with the kit supplied. However, most building merchants stock a vast array of pipework, so either pull off an existing connector (most rainwater pipes are push-fit) and take that with you or measure the external diameter of the pipe and try to find one of a suitable size. Square pipework is fairly easy to measure, but to measure the diameter of a round pipe you'll either need a set of callipers (and few of us have these lying around the house) or a piece of string and a ruler. The great thing about the string and ruler approach is that after years of boring maths lessons you finally get a chance to use pi in anger:

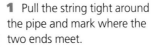

1 Pull the string tight around the pipe and mark where the two ends meet.

2 Measure this distance.

3 Divide this measurement by 3.14 (pi). The resultant number is the external diameter of the pipe.

You need to ensure that the inlet and overflow pipes are roughly the same size. If not there's a danger of the butt overflowing when it rains hard. Most kits get around this by supplying an inlet that only collects a small percentage of the water coming down the pipe. There's a temptation to try and collect all the water, especially if you have a large butt and are taking the water off a fairly small area, but if you do this you're going to have to adjust the size of the overflow pipe accordingly.

10 IN THE KITCHEN

Fitting new monobloc taps   119

Fitting a new kitchen sink   121

**Fitting a new dishwasher or
washing machine**   124

Fitting a fridge ice-maker   128

Fitting a water filter   129

Fitting a water softener   129

Fitting new monobloc taps

Fitting new kitchen taps is easy. Getting the old taps out is the hard part.

Most modern kitchen taps are what are called monobloc taps, consisting of a single block of brass with a hot and cold tap going into it and a single spout coming out, and that's the kind we're going to look at here. If you have traditional separate hot and cold taps then pop over to 'Replacing washbasin and bath taps' in Chapter 11.

Tools and materials

- Set of monobloc box spanners
- Pipe-slice or hacksaw
- Adjustable spanner
- Pump pliers

1 To fit a monobloc tap you'll need a set of box spanners. These go by many names but if you ask for a 'monobloc, back nut, box spanner set' you won't go too far wrong.

2 Turn off the hot and cold water. Ideally you'll be able to do this by turning off the isolation valves under the kitchen sink, but if there aren't any consider fitting them (see earlier chapters). With the water off, open the kitchen tap to check and to let the water drain away.

TIP *There are often drain cocks under the kitchen sink. Once you've closed off the water and opened up all the taps, open these and catch the water in a bowl. When the water stops flowing tighten them up again. Doing this takes any remaining water below the level you're going to be working at. Don't do this if you've turned off the water to the taps via isolation valves!*

If you've turned the water off for the whole house you need to open up all the taps and flush all the toilets to drain the system. Water is very sticky stuff and will just hang, suspended, in the pipes whilst all the taps are closed. However, the moment a tap is opened or a toilet is flushed the suspended water will race down the pipework to find you stuck under a sink...

3 Disconnect the tap from the hot and cold pipework, either by using a pipe cutter to slice through the copper or by undoing the closest compression fitting to the tap.

⚠ WARNING

Often the closest compression fitting is the isolation valve. Whilst it's perfectly safe to loosen off the nut at the tap side, you'll get very wet very quickly if you loosen off the nut on the water side. This sounds very obvious, because it is, but there are a lot of damp people out there who made this very mistake. Regardless of which end you undo, always hold the valve steady with a set of pump pliers whilst you undo the nut.

4 Take your new box spanner set and crawl under the kitchen sink. Above you, you'll see the base of the tap, with a single nut holding it in place. Find the spanner that fits this nut and turn it anticlockwise to remove it – you might want to ask someone to hold the tap for you.

5 Now remove the metal backplate and the rubber washer underneath this.

6 You should now be able to remove the tap by lifting it up. If you can't get the pipework through the hole you can twist the pipe anticlockwise to unthread it from the base of the tap or cut off the nuts at the end using a pipe-slice or junior hacksaw. If the tap

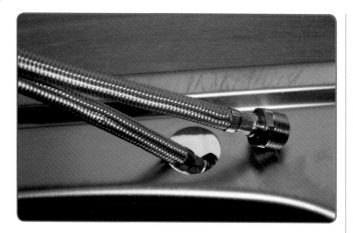

is connected via flexi tap connectors then just pull one connector through the hole at a time.

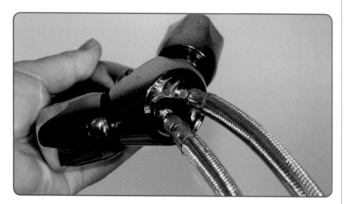

7 Your new tap will come with either two copper tubes or two flexible connectors that screw into its base. Hand-tighten these to the base of the tap, making sure that you know which is for hot and which is for cold – some manufacturers mark the connectors, red for hot and blue for cold. Once you know which is which, make a mark on at least one pipe so that you'll still know once you're back under the sink.

TIP *If your tap came with copper tails you should really consider replacing them with flexible connectors, as it's really going to save you a heap of time and trouble. The size of the thread at the base of your tap varies between 10mm and 12mm, so either measure them or, easier still, take one of the copper tubes to your local plumbing merchants and ask for 'a flexi one of these, please'.*

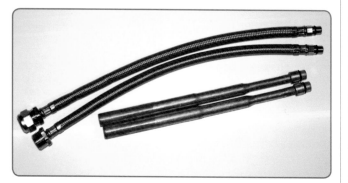

These connectors only need to be hand-tight, but if in doubt give them the tiniest of tweaks with an adjustable spanner.

If you choose to persevere with the copper tails you can gently bend them into position by hand, but try not to shorten them with a pipe slice as the copper is usually so thin that the tube warps.

8 The tap is held to the kitchen sink by a threaded bar that connects to a backplate and nut underneath the sink. You need to fit this bar at this stage. It usually has a groove in one end so you can use a screwdriver to screw it into the base of the tap.

9 Most taps come with some sort of base arrangement. Often it's just a rubber washer that covers the base of the tap, but it can be more elaborate. Fit these at this stage.

10 Now thread the pipework through the hole and check that the tap is sitting properly on your sink. It's a good idea to ask someone to hold the tap still whilst you work on it. Now check that the tap is the right way around – the cold should always be on the right, the hot on the left. To remember this I always think about the UK weather – right cold!

11 The rest is now the reverse of what you did to get the old tap out, so follow steps 5–2 in that order. More often than not you can use the old pipework to connect the new tap to the water supply. For more details on using soldered joints or compression fittings read Chapter 6.

TIP *A common mistake is to bend the flexi connectors too tightly. This just restricts the flow of water, rendering the tap worse than useless. So keep all bends gentle and don't let the flexi-hose twist when you're tightening the connector.*

12 Close the hot and cold on your newly installed tap and turn the water back on. Check everything for leaks and then open the taps and check all is working properly.

Leaks can take hours to develop, so rather than just giving your work a quick glance and then refilling the kitchen-sink cupboard with hundreds of boxes and bottles, leave the cupboard empty for the day and recheck your work in the evening or the next morning, so that you're certain everything is still dry.

Fitting a new kitchen sink

If you're replacing an old sink remember that you need a new one that's either the same size or bigger. If you opt for a smaller replacement then you're going to have to replace the entire worktop. The following steps assume you're fitting a sink from scratch but the process is pretty much the same if you're replacing an old sink.

Tools and materials
- Masking tape
- Pencil/marker pen
- Drill bit size 10–14
- Drill
- Jigsaw with downward-cutting blade
- Clear sealant
- Pozi screwdriver
- Large flat-head screwdriver or old chisel

1 Decide on a location – the bowl needs to be inside a kitchen unit, ideally in the centre so you can get access to it. Don't place it too far back or you'll struggle with fitting the taps; too far forward and you're past the front of the cupboard and into the lip of the worktop.

The new sink will often come with a paper template that can be stuck to the worktop with masking tape. If this is the case, stick it down, check it's aligned with the worktop properly and then jump to step 6. Often the box the sink comes in has the template marked on it, so check this before you throw the box away.

If you're devoid of templates turn your sink upside down (no taps should be fitted to it at this stage) and lay it roughly in position on the worktop.

2 Put a bit of masking tape under each corner and mark the edge of the sink.

3 Remove the sink and use masking tape to connect your marks together, making sure you cover an area at least 4cm wide.

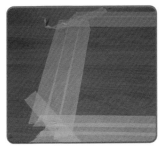

4 Put the upside-down sink back into position. Measure the front edge at both ends to make sure it's aligned with the worktop. Once you're sure it's lined up properly use a pencil or marker pen to draw around the edge on to the masking tape.

5 Measure 1cm in from this line and use a straight edge to connect up this inner line. This is the line you're going to cut, so it's a good idea to mark it in a different colour and to score the outer line so you don't follow that by accident.

6 Drill a hole just inside the inner line using a size 10–14 drill bit.

7 Fit a downward-cutting wood blade to your jigsaw. The downward-cutting bit is important to stop the blade chipping the laminate worktop surface – if your worktop isn't a laminate then this isn't such an issue.

The reason you want a track of masking tape at least 4cm wide around your cut hole is so that the worktop surface is protected from damage as you push the jigsaw around it. An alternative is to buy a jigsaw with a smooth plastic 'shoe' – although I still put masking tape underneath as well.

Put the blade into the hole you've just drilled. Check all is clear underneath the area you're about to cut and start cutting along the inner line you marked. Ask someone to support the area you've cut out as you approach the end.

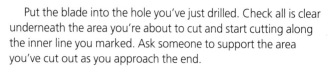

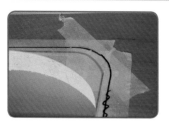

TIP *To cut rounded edges it's often easier to cut the hole out of the worktop first and then go back and tidy up the edges afterwards when the jigsaw blade has more room to turn. Alternately you can buy very thin blades that are designed for cutting tight corners.*

8 Fit the sink into the hole to check it fits OK. If it doesn't, have a look underneath to see where it's catching and make any adjustments as necessary.

9 You now need to seal the hole you've just cut so that if any water escapes from the sink it won't damage the worktop. The easiest way to do this is to run a wavy line of clear sealant around the fresh cut and use a piece of cardboard to smear it over the cut line.

This step is particularly important for laminate worktops with chipboard or MDF inside, as the surface will blister if it gets wet.

10 While the sealant dries fit the foam edging strip that usually comes with the sink. If your sink didn't come with one you need to apply clear sealant to the sink edge – only do this just before you fit the sink.

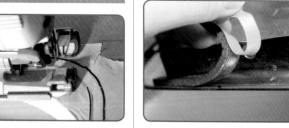

11 To fit the tap to the sink follow either the 'Fitting new monobloc taps' directions earlier or 'Replacing washbasin and bath taps' in Chapter 11, remembering to peel back any protective covering on the sink before fitting the taps.

Most sinks come with the tap hole already cut. Most actually come with two holes, so you can have the drainer to either the left or the right of the bowl. Decide which hole you want to use and use the little blanking plate supplied with the sink to blank off the other one.

If you have a 'stone effect' sink the hole isn't always pre-cut. Instead they half-cut through it and leave it to you to finish the job. To do this you need a hole-cutter – and nerve.

12 The sink will usually come with lots of little metal clips. If this is the case clip these to the sink at this stage, after first checking the hole you've cut to see if the clips can actually be reached for tightening up later. Ideally you need clips on either side of an edge and in the middle. The sink edge will have little holes cut into it to accept these clips.

The clips just push into position – but be careful, as there are lots of sharp bits of metal involved at this stage.

13 Now remove the remains of the protective covering on the sink and the masking tape around your hole and lower the sink into position. You'll have to push the clips in a bit to get them through the hole, and they have an annoying habit of springing free at this point.

14 Have a friend hold the sink down while you crawl underneath and tighten the clips. To do this you first need to push the 'teeth' of the clip over the base of the worktop, then tighten the screw to draw the teeth into the base. You just want to draw the sink down and keep it secure, so don't overdo it or the clips will just pop out.

15 With the sink in place you can now fit the tap to the hot and cold pipework.

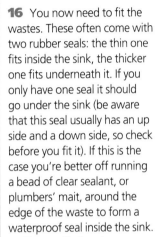

16 You now need to fit the wastes. These often come with two rubber seals: the thin one fits inside the sink, the thicker one fits underneath it. If you only have one seal it should go under the sink (be aware that this seal usually has an up side and a down side, so check before you fit it). If this is the case you're better off running a bead of clear sealant, or plumbers' mait, around the edge of the waste to form a waterproof seal inside the sink.

There will often be an overflow fitting and a connector on the base of the waste. The overflow will usually come with a single washer that fits on the outside of the sink. The flexi connector between the overflow and the waste often needs cutting back to avoid any U-bends – these will only fill with gunk over time.

17 To tighten the waste into position there's a rather thick threaded bar or screw. Push this through the middle of the waste and hand-tighten. The groove on this is usually too big to tighten fully with a screwdriver so I use an old chisel to get them nice and tight.

Remember you're tightening the waste into a plastic base, so don't overdo it or you'll just crack the base.

18 You now need to fit the waste pipework. The manufacturer usually supplies this and fitting it is a bit like taking part in the Krypton factor. Each kit is different, so we won't go through the procedure, but here are a few tips:

- Use a plastic pipe cutter to cut back pipework where necessary – although most is designed to slide inside each other so that you can adjust it without cutting.

- If your waste comes with washing machine connectors, open these up before connecting anything to them as they're usually blocked off inside by a blue plug. If you forget to do

this your washing machine and dishwasher will stop working as soon as they reach the 'drain' part of their program. If this happens check for these blue plugs again.

- If you're not going to use the washing machine connectors check that this blue plug is in place before you test the waste.

- See Chapter 6 for how to use waste fittings. Don't just push the fittings together, dismantle them first.

- Make sure you use the right size pipework – 40mm or 1½in for kitchen-sink wastes.

- Only the final U-bend of the manufacturer-supplied pipework will generally fit directly on to a standard waste pipe. If you need to fit standard pipework at any other point you'll need to connect the pipes using a 40mm 'universal compression' fitting.

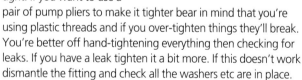

- Waste pipework usually only needs to be hand-tight. If you want to use a pair of pump pliers to make it tighter bear in mind that you're using plastic threads and if you over-tighten things they'll break. You're better off hand-tightening everything then checking for leaks. If you have a leak tighten it a bit more. If this doesn't work dismantle the fitting and check all the washers etc are in place.

- The water has to move downhill at all times. Don't be tempted to ignore this fact, as doing so might work for a week or two but you'll invariably get leaks and blocked pipework in the end.

19 With everything tight, turn the water back on and check for leaks in your hot and cold pipework.

To test the waste, first just run water into the sink with the plug open. If it passes this test, close the plug and fill the bowl with about an inch of water. Now open the plug and check that all is still dry.

If there is an overflow on the sink then fill the sink till it reaches the overflow and check it works without leaking.

Fitting a new dishwasher or washing machine

Most dishwashers and washing machines only need a cold-water feed and access to a waste pipe.

For the cold water you'll need to fit a washing-machine connector to the cold-water pipework. This can either be at the end of a pipe run or you can buy a tee connector and fit it in the middle of a run.

See 'Using a compression fitting' in Chapter 6 regarding how to fit the connector. This connector is there to isolate the machine from the water so that it can be repaired or removed – when the tap's turned across the flow it's closed, when it's in line with the pipework it's open.

Open

Closed

The easiest way to deal with the waste is by connecting the hose to a waste connector directly under the kitchen sink. Most sink wastes now come with these adapters, which are blocked off internally by a blue plug when not in use. To use one of these adapters:

Tools and materials

- Screwdriver
- Jubilee clip

1 Remove the spout by undoing the nut.

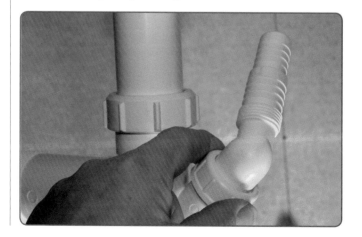

2 Remove the plug. These plugs vary, but if it's there at all it's usually blue. The main aim here is to check that the adapter hasn't been blocked off.

3 Check the end of the spout is open – some manufacturers don't supply a plug, instead the end of the spout is solid and you have to cut it off to reveal the hole – and check it fits on to your washing machine hose.

4 Push the spout back on, making sure that any washers are in place – most don't come with washers these days – and hand-tighten the nut.

5 Pass a jubilee clip over the end of the hose and push the hose

on to the spout. Once in place tighten the jubilee clip using a screwdriver. Remember that you can buy longer hoses if necessary.

If your sink doesn't have any appliance connectors, don't worry, you can always buy a connector that can be added to the waste pipework already there.

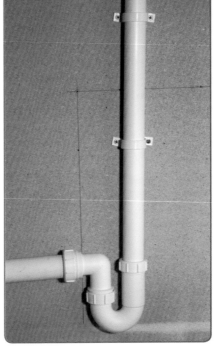

If you can't use the waste under the sink you can buy a stand-pipe, which is an open-ended individual trap. To use one you simply push the hose from the dishwasher/washing machine into the top of

the stand-pipe, though it's better to attach the hose to a plastic bend support – this stops the pipe kinking, and the support can be screwed to the wall to ensure the hose never falls out of the stand-pipe. The stand-pipe itself should be connected to 40mm pipework (no smaller, and if there are a lot of things using this piece of pipe you might want to go up to 50mm), and usually runs out through an external wall and into a gully.

If you're going out through an external wall hire a 'core' drill set to make a nice neat hole – most builders' merchants and hire shops supply these. You'll want a core drill about 50mm in diameter for

40mm pipe. You'll need to use either mortar or clear sealant to create a draught-proof and waterproof seal around the pipework afterwards.

When locating everything bear in mind that the waste is pumped out of the machine. This means that the adapter or stand-pipe can be set higher than the machine – there is a height limit, so check the manufacturer's manual first. However, from the adapter/stand-pipe it's gravity that does all the work, so you must have a downhill run into the main waste.

DISCONNECTING A DISHWASHER OR WASHING MACHINE

When you order a new washing machine or dishwasher the company supplying the replacement will often take away your old one. Sadly, there's often a proviso: they won't disconnect the old one from the water supply and they won't fit the new one. You could call out a plumber to take care of this but it's a lot cheaper to do it yourself.

Tools and materials
- Pump pliers
- Bowl or sponge
- Screwdriver

1 Carefully pull the washing machine/ dishwasher out. As you do so you'll see two or more hoses coming from the back and connecting to the water supply. There is also a hose going to the waste pipework, and an electric cable. Unplug the cable and carefully pull the appliance out so that you can get to the hoses.

2 The reason that the people delivering the new appliance won't disconnect the old one is because there's a danger of flooding if you disconnect the appliance and the washing machine connector is damaged and doesn't close fully.

With this in mind turn off the water in the house – hot and cold if your appliance has two hoses. I wouldn't drain down the system at this stage, because whilst you don't want a flood you do want to know if the valves are failing – because you'll need to replace them if they are.

3 You can now see the washing machine tap connectors. These are usually chrome isolation valves with a ¾in thread on one side that attaches to the appliance's hose. In the middle they have a little plastic lever, usually blue for cold water and red for hot. To turn the valve off you need to turn the lever so that it's pointing at right angles to the pipework coming into the valve.

Open

Closed

4 Put a bowl or a sponge under the valve and undo the hose. You can usually undo this connector by hand, but use a pair of pump pliers if need be. Some water will come out of the hose pipes and this will increase if you lower the hoses to the floor, so best empty them into your bowl.

5 Check that the isolation valves are closed and aren't leaking. If all is dry turn your water supply back on and check again. If they're leaking follow these steps:

- If you haven't already done so, read 'Using a compression fitting' in Chapter 6.
- Turn the water off again and drain the system at the kitchen sink.
- Open the faulty valve to drain the remaining water out of it.
- Remove it at the back nut.

Fit a new valve, reusing the old back nut and olive – if the olive looks damaged either use PTFE tape over it or think about cutting the pipe back using a pipe-slice and fitting a new nut and olive on to fresh pipework.

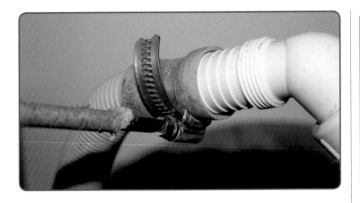

6 The waste hose will either empty into a stand-pipe, in which case just pull it out and lower it to drain the excess water into a bowl, or it will connect to the waste pipework under the kitchen sink. If confronted by the latter scenario you need to loosen off the jubilee clip at the end of the hose and then pull the hose off the connector.

If you're not going to be replacing the washing machine/dishwasher immediately you'll need to blank off this connector – pop down to your local plumbing merchants and they'll usually have something that will work. To fit it, take the nut off the original adapter and use this to secure the plug.

REMOVING THE HOT-WATER CONNECTOR
When your new appliance arrives have a look to see how many hoses it has. A lot of older appliances needed hot- and cold-water connections, whereas most new ones only need cold. You could just leave the hot-water isolation valve turned off, but it's better to cap off the hot water altogether if it's not going to be needed. To do this:

Tools and materials
■ Pipe-slice
■ Push-fit stop-end

1 Turn off the hot water and open the nearest tap to your appliance. When the water stops running put a bowl under the hot-water washing machine connector and open it; some water should come out, but not much.

2 Use a pipe cutter to cut off the old connector and leave a piece of clean, straight, copper or plastic tube.

3 Push a speed-fit stop-end on to the pipe to permanently isolate the water supply. If you have plastic pipe make sure you use a plastic pipe insert first – see 'Plastic push-fit pipework' in Chapter 6 for more details.

4 Turn the hot water back on – leave that one tap you opened earlier in the open position until the hot water is flowing through it properly and all air has been expelled from the tap.

5 Check that all is dry.

FITTING THE NEW APPLIANCE
Having followed the earlier steps you should know where your washing machine or dishwasher connects to the water supply and that the isolation valve that serves this connection works properly. To fit the new appliance to the water supply:

Tools and materials
■ Pump pliers
■ Jubilee clip
■ Utility knife

1 Usually the hoses are already attached to the appliance, but if this isn't the case attach them now, making sure that any rubber washers supplied are in place – they're usually fitted into the hose itself and don't come out.

2 Take the other end of the hose and screw it on to the washing machine connector. Bear in mind that you're threading plastic on to brass, so only turn it by hand – if you can't, you probably don't have the hose aligned right, so undo it and start again. Once

the connection is hand-tight give it a quarter-turn using a pair of pump pliers.

3 Open the washing machine connector valve by turning the plastic lever so it's in line with the pipework.

4 Check there are no leaks.

5 Fit the waste hose either into a stand-pipe or a connector under your kitchen sink.

For the stand-pipe you need to ensure the hose is far enough down the pipework to stop it falling out but not so far that it's blocking the U-bend at the bottom. It's always best to attach the hose to a support and then screw the support to the wall.

For the connectors under the sink you need to open them up and check that they're not currently blocked – usually it's a little blue plug. Once you've verified that they're open push the hose on to the connector (you might need to cut the connector back a little to get a good fit) and use a jubilee clip to stop the hose falling off.

6 You're now ready to push the appliance back into position. When doing so, be careful not to trap or bend the hoses too sharply.

Fitting a fridge ice-maker

The current trend seems to involve furnishing your kitchen with a fridge roughly the size of a garden shed. In addition to being able to house a year's worth of canned beer these fridges often come with an ice-making machine. The instructions on how to fit it are usually written in hieroglyphs specifically designed to confuse everyone within a matter of minutes.

Every ice-maker is slightly different but the basics are as follows:
- Invariably the connection to your mains water is via a proprietary connector and a 6mm length of plastic hose. 6mm isn't generally used in plumbing so if you lose or damage the hose or the connector you may well struggle to find a replacement – try a company dealing with refrigeration first, as they tend to use pipes of this size.
- Always connect to the cold water. This sounds obvious but I've come across them connected to the hot water!
- If you have a water softener in your home make sure you connect the ice-maker to the *unsoftened* water supply – see 'Fitting a water softener' later for more details.
- If you have any doubts, read Chapter 6 on how to use push-fit and compression fittings.

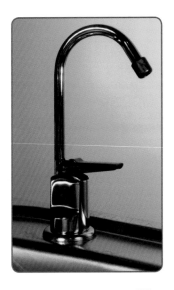

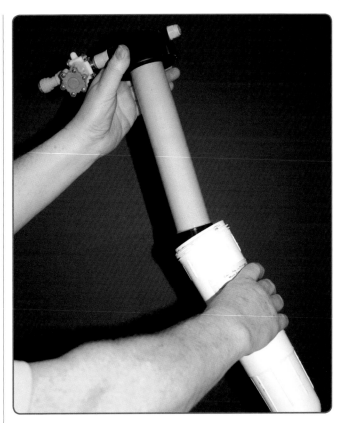

Fitting a water filter

If you have lead pipework inside your home, or if the mains pipe coming into your home is lead, then it's a really good idea to fit a water filter to at least one tap and only take drinking water from this. The vast majority of water filters will filter out heavy metals, but check this before you buy.

There are two main ways to use a water filter: you can fit a separate little tap to your sink, or you can fit a 'tri' tap – one that supplies hot water, normal cold and filtered cold water. The latter type looks much better but is far more expensive.

How you fit the water filter will vary from manufacturer to manufacturer but the principle is very simple: you connect the inlet of the filter to the cold-water mains and you connect the outlet to a tap. The following are some basic tips to help you along the way:

- Connect it to the cold-water supply. If you have a water softener make sure you connect the filter to the *unsoftened* water supply.
- Read Chapter 6 regarding how to use push-fit and compression fittings.
- Read 'Fitting new monobloc taps' at the beginning of this chapter for the basics on fitting a tri-tap.

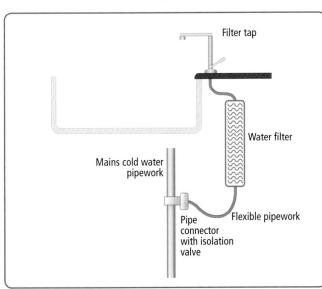

Mains cold water pipework — Water filter — Flexible pipework — Filter tap — Pipe connector with isolation valve

A water filter isn't a water softener. It will usually leave the 'hard' minerals in the water – because these are good for you – but remove things like lead and fluoride, because these are either bad for you or give the water a nasty taste. A water softener does the opposite; the lead will still be there but the good minerals will have gone. As such, water filters are good for you, water softeners are good for showers, washing machines, dishwashers etc.

The filter itself will wear out over time – generally they need replacing every 12 months. Some will give you a warning when they need replacing, for others you'll need to jot down a reminder in your diary. As a general rule, to change the filter you just untwist the main container then remove and replace the filter within.

Fitting a water softener

Most areas of the UK have what's referred to as 'hard' water. This means that the water contains minerals that sediment out when the water's heated, coating pipework and ruining your white appliances, showers, combi boilers etc. Whilst this is bad for your appliances it's not bad for you, in fact hard water is much better for you than soft water – the minerals that damage your white appliances are the same minerals that bottled water suppliers rave on about. So when you fit a softener you

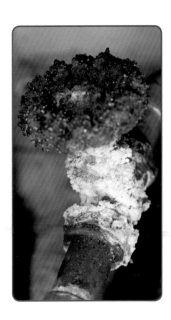

need to ensure that at least one tap is still supplying drinking water, *ie* you don't want to soften all the water in your home.

TIP *To find out how hard your water is go to the website of your local water company, search the site for 'water hardness' and enter your postcode. Most water companies offer this service.*

To muddy the waters a bit there are lots of different devices that claim to 'soften' water:

ELECTRIC OR MAGNETIC WATER CONDITIONERS

These don't 'soften' the water, in that the chemical composition of the water isn't changed one iota. What they do claim to do is 'physically condition' the water by passing it though a magnetic or electrical field. The idea is that this alters the minerals in such a way as to prevent them coating the pipework any more. Personally I think you're better off sacrificing a chicken to the God of Limescale, but some people swear by these devices and under laboratory conditions there is an 'effect', often dependent on the amount of iron and zinc present in the water at the time.

The benefits of such devices are that they're cheap and very easy to fit. They also don't interfere with things like boiler filling loops or water filters. The downside is that they can stop working over time with little or no warning – always assuming that they had any effect in the first place.

ZINC WATER CONDITIONERS

These usually look exactly the same as the electric/magnetic conditioners and often come with electric or magnetic fields. The idea is that they release zinc into the water supply. This zinc then binds to the 'hard' minerals, stopping them from coating the pipework. The problem is rather ironic, in that the zinc source often gets covered in limescale, stopping the release of zinc and thus stopping the device from working.

These tend to be slightly more expensive and have the same pros and cons as electric/magnetic conditioners.

PHOSPHATE CONDITIONERS

These work the same way as the water softening tablets available for your washing machine/dishwasher. They use trace amounts of phosphate, rather than zinc, to condition the water. Again, the idea is that the resultant water is less likely to deposit limescale inside pipework.

These are a bit more expensive but can generally be proven to work in the real world. They need to be topped up periodically with 'food

quality' phosphate, and can stop working altogether if the dosing holes in the mechanism get blocked.

ION REPLACEMENT WATER SOFTENER

These are the real McCoy and are the only ones that actually alter the chemical composition of your water so that limescale won't form. The downsides are that they're pretty expensive, quite bulky and use salt to regenerate the ion-exchanger. This salt can theoretically get into the water supply, so it's not recommended that you drink the softened water. Also, because softened water is more acidic it's not recommended that you fill your CH system with it – although the hot water supplied by a combi is better off softened, *ie* soften the cold water going into your combi.

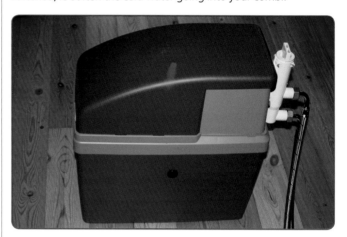

Most water softeners work the same way, so we'll have a brief look at the installation process (the connections they use are essentially the same as those of a dishwasher/washing machine, so you should also take a look at the earlier section on 'Fitting a new dishwasher or washing machine'):

1 You need to intercept the water as soon as it comes into the home if you're going to soften the vast majority of it. At the same time you need to be quite close to a tap that can be left unsoftened so that it remains drinkable. Generally speaking most units go under a kitchen unit close to the mains stop tap.

2 Decide what kind of pipework and fittings you're going to use (plastic, copper etc) because you'll have to alter the cold-water pipework. You need to arrange it so that unsoftened cold water is supplied to the kitchen tap, any outside taps and any central-heating filling loop. Also, you need to be able to bypass the

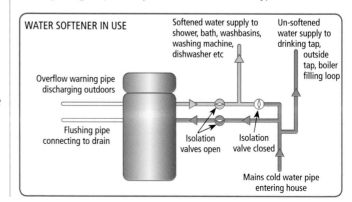

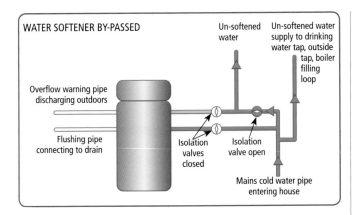

WATER SOFTENER BY-PASSED

Overflow warning pipe discharging outdoors

Flushing pipe connecting to drain

Isolation valves closed

Isolation valve open

Un-softened water

Un-softened water supply to drinking water tap, outside tap, boiler filling loop

Mains cold water pipe entering house

water softener in case it needs to be removed or isolated – some softeners have a bypass built into them. Bypassing is usually achieved by connecting the unsoftened cold-water pipework to the softened water pipework via an isolation valve that can be left closed when the softener's in use.

TIP *Supplying unsoftened water to the filling loop on your CH system is usually an expensive, tiresome, if not impossible task. You may be better off just isolating the water softener, running 'normal', ie unsoftened, water through a tap close to the filling loop to flush the softened water out of the pipework and then using the filling loop. When you've done that just open and close the appropriate isolation valves to reconnect the softener to your cold water.*

3 Connect the inlet and outlet hoses from the water softener to the appropriate pipework using washing machine connectors.

4 When the softener needs to regenerate the ion exchanger it will flush lots of salt water through itself. This then needs to be discharged to a drain. The easiest way to achieve this is to connect the waste outlet hose to a washing machine waste connector under the kitchen sink – remembering to remove any plug from this connector first.

5 If anything goes wrong when the softener is regenerating you need to make sure the water escapes from an overflow pipe rather than flooding the kitchen. To achieve this run 19mm overflow pipe direct to the outside and terminate it just above the ground. Remember this pipe is to warn you that things have gone amiss, so you need to be able to see that water's coming out of it. At the same time you don't want this water to do any damage to the house.

6 Most softeners need to be connected to an electrical supply. Fortunately this is usually just a matter of plugging them in. To make things even easier some manufacturers now make softeners that require no electrical supply at all.

7 Fill the bucket at the front of the softener with salt – dishwasher salt will do, although you can buy bigger bags for water softeners. Do not use rock salt! There's usually a line painted on the inside of the unit to show how high the salt level needs to be kept.

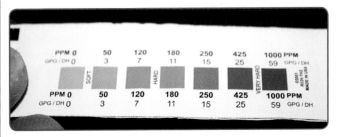

8 Many softeners will automatically adjust the number of times they regenerate based on the hardness of the water and the amount of water being used. To do this they need to be told how hard the water is, and most manufacturers provide a test pack to determine this. The pack usually consists of a test tube and tablets that you add until the water changes colour. Other companies supply paper strips that you just dip in your water and check the colour change.

Often the value you need to enter is in 'parts per million', whilst the test kit invariably gives a value in 'degrees Clark'. To convert, treat one degree Clark as 14.25ppm, or 1ppm as 0.07 degrees Clark.

If this all sounds too much just go to the website of your local water company and look up how hard the water is in your area.

9 After a few days' use the test kit the manufacturer gave you and see how hard the water is now you're softening it – remember not to test on the kitchen tap as you're probably not softening that one. Repeat this a month later to test that the system is regenerating itself properly.

10 Keep the unit topped up with salt. Some machines will 'beep' when they need more salt, others just need topping up on a regular basis.

11 Rave on to all who'll listen about how efficient and long-lived your white appliances have become … Alternatively, just luxuriate in the wonder of a shower of softened water – it's amazing the difference it makes.

11 IN THE BATHROOM

Home Plumbing Manual

Using sealant   133

A very brief guide to tiling   134

Replacing washbasin and bath taps   135

Removing an old toilet   137

Fitting a new toilet   139

Removing a bath   146

Fitting a new bath   147

Removing a washbasin   157

Fitting a new washbasin   158

Fitting a new shower   164

Fitting a bidet   169

Fitting a shower tray and cubicle   170

Swapping your bath for a shower tray   171

Wet rooms   171

This is where the majority of plumbing-related work takes place and it's often the first major project undertaken by the keen DIYer. If you've read Chapter 6 and had a bit of a practice then nothing here should hold any terrors – the steps given assume that you *have* read Chapter 6 and practised a bit. My only real word of advice relates to tiling. It's what everyone's going to notice first, so if you've never done it before read the next page and then read up about it in detail elsewhere. You could, of course, just chuck the tiles on the wall and claim you're aiming for a 'rustic' look, but I really wouldn't rate your chances.

If you're intending on buying a new bathroom suite bear the following in mind:

- There's invariably a very good reason why a suite is cheap. A good suite should last at least ten years, so it's worth buying the best you can afford.
- The descriptive term 'white' covers a wide variety of colours. Just because a toilet from one range claims to be white doesn't mean that the washbasin from another range is going to be the same white, so it's usually best to buy a suite; it's usually cheaper as well.
- The taps, or brassware, can double the price of a suite, so always check to see if the price includes or excludes brassware.
- 'Designer' suites are invariably a nightmare to fit and service. Check out the installation process before you buy.
- Suites bought-in from overseas often come with non-UK standard connectors. Before you start installing, check the connectors provided and, if need be, pop down to a plumbing merchant and buy the necessary adapters.

Using sealant

If you're going to fit a bathroom you're going to have to learn to use silicon sealant. It's actually very easy to use, but this hasn't stopped people making a right hash of it. First off, to make life easy, buy a sealant spreader or 'profile'.

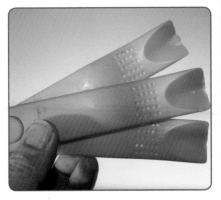

1 Cut off the lid of the sealant tube.

2 Put on the nozzle and cut this to size – a hole 3-5mm in diameter is about right.

3 Fit the sealant tube into the gun.

4 Work the sealant to the top of the nozzle and then apply a single long string of it to whatever edge you're sealing. Keep the nozzle at a slight angle whilst applying an even squeeze to the gun trigger. At this stage we're just looking to get a decent amount in place, so wilst we're aiming to keep things tidy it's not essential at this stage.

5 Now take your profile tool and run it forward over the line of sealant. As you go it removes all excess and, if you've applied enough, it should leave a nice smooth line behind it.

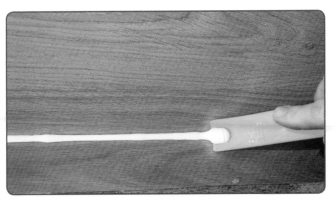

A (very) brief guide to tiling

Since we've already looked at kitchens and we're now going on to look at bathrooms, it seems appropriate to touch on the dos and don'ts of tiling.

We live in a world of twisted priorities and contorted values. As such the vast majority of people won't care how fantastic your plumbing work is if the tiling looks rubbish. Tiling is the finishing touch to most major kitchen and bathroom projects so be sure you know what you're doing before ruining some perfectly good plumbing work. Haynes publish a manual on this very subject, so buy it.

Some manufacturers and members of the frilly-cuffs brigade will try to persuade you that porcelain tiles are wondrous. Don't listen to them; porcelain tiles are horrendously hard, so much so that working with them is far more difficult than with standard ceramic tiles. Unless you intend to spend your time tap dancing around your bathroom with steel lined clogs I'd avoid porcelain. Exceptions are things like kitchen floors, hallways and nuclear bunkers, where the robustness of porcelain can be a genuine bonus.

The same guys who persuaded you to look at porcelain tiles will also tell you that natural stone tiles are the ultimate in decadent living. Natural stone tiles can look awesome in the right location but they usually need treating before they're fit to use in a bathroom or kitchen. Even then they can start to let water through in very damp areas – shower cubicles, for example – and can become home to a wide variety of moulds that enjoy the nooks and crannies of a natural stone finish. Such tiles can also be very soft compared to ceramics. Have I put you off using them in

bathrooms? If the answer is 'No' go back to the top of this page and keep reading it until I have.

Baths and, to a far lesser extent, shower trays can flex when they're in use. Bear this in mind when tiling up to the edge of a bath or shower tray. You should aim to leave a 3mm gap between the edge of the tile and the bath/shower tray. When you've finished tiling, fill this gap with sealant and not grout.

The most important rule of tiling is too keep things clean as you go along. Removing damp tile adhesive and damp grout is a piece of cake; removing dried-on adhesive and grout is a nightmare.

Remember that you're not in a race, no matter what your partner might be suggesting. Once done you can expect your tiling to last at least ten years, so take the time to do it right.

In a bathroom I always fit the bath and the shower tray before doing the tiling. Once the tiling is finished I then fit everything else. The reason for this is simple: the bath and tray are less likely to leak, it creates a cleaner finish and it allows you to change the toilet and washbasin without having to retile.

Most tile designs have a fairly short shelf life and can change colour from batch to batch. As such it's a good idea to have a box left over to cover yourself for future changes or breakages. Most retailers will let you take back any unused boxes so it's always better to buy too many rather than too few.

An alternative to tiles is waterproof panels. These can look fantastic when used well but can make a bathroom look a bit clinical if you're not careful about the colour choice and the amount you use. Some people insist that they're easier to fit than tiles. Personally I'm not so sure.

Done right, tiles can look awesome.

Natural stone can be filled with nooks and crannies.

Waterproof panels are a good alternative.

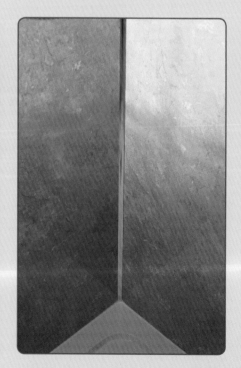

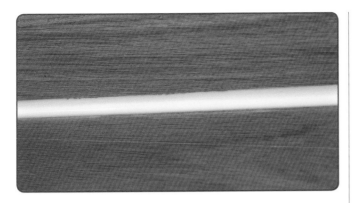

6 That should have done it, but if there are any blemishes apply a little more sealant and run the profile over these areas again. You can finish off by lightly running a dampened finger over the sealant line.

7 If you've got sealant where it shouldn't be just wipe it off with dry tissue paper or use the profile tool to remove it.

TIP *For the corners you might find it easier to use the tool to remove the excess sealant and then a dampened finger to get a smooth finish.*

Once the sealant starts to dry it's best to let it dry completely before attempting to modify any blemishes, otherwise it just starts to cut up and look horrible.

It's always a good idea to have lots and lots of tissue paper on hand when using sealant.

Replacing washbasin and bath taps

As with most taps the only real difficulty is the location – they're just plain awkward to get at. This is the reason they invented the basin wrench, and without one you've as much chance of removing a basin tap as you have of plaiting jelly. The other tool you'll need is a plumbers' ½in and ¾in box spanner.

The basin wrench works by setting the head at right angles to the handle – you can offset it if needs be. To do this just grab the head and turn it. If you turn it one way it'll tighten a nut; if you turn it the opposite way it'll loosen it.

Once set just slip it over the nut and twist. If you've set it the wrong way it'll just slip around the nut; if you've set it right it'll tighten as you twist it and away you go.

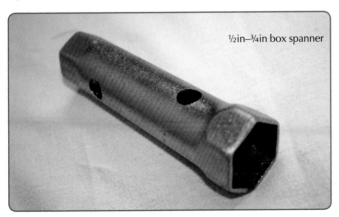

½in–¾in box spanner

Tools and materials

- ■ Basin wrench
- ■ ½in–¾in box spanner
- ■ Replacement fibre washers
- ■ A friend

1 Turn the water off either at the isolation valves or for the whole house. Open the taps to make sure they're off and open a downstairs hot- and cold-water tap to drain the water out of the pipework.

2 Set your basin wrench to loosen nuts and lie under the basin or bath so you can see the tap connectors at the end of the tap tails.

3 Slip the wrench over the nut and turn it anticlockwise to loosen the connector.

4 Once loose you should be able to pull the pipework away from the tail of the tap.

5 Now use the box spanner to loosen off the nut at the base of the tap. Usually the tap will turn at

this point so you need someone to hold it steady. To increase the power you can apply you might have to put a screwdriver through the box spanner.

6 You should now be able to lift the tap out.

7 If the tap connector uses fibre washers it's a good idea to replace them at this stage. Remove them using a small

screwdriver and make sure all trace of the old washers has been removed before fitting the replacements.

8 Fit new washers – first checking the connector is free of any debris.

9 The new taps usually come with a rubber washer that fits under the base. Push this on and then put the tap in place.

10 Fit the back nut to the base of the tap, set the

tap to the right position and use the box spanner to tighten it until the tap no longer moves – again, it's best to have a friend on hand to hold the tap steady for you.

11 Double-checking that your new washer is still in place, move the pipework back under the tap and hand-tighten. It's best to push the washer on to the base of the tap and then start to tighten the nut.

12 Finish off by giving it a three-quarter turn with the basin wrench.

13 Turn the taps off, turn your water back on and check for leaks.

TIP *For basin monobloc taps see 'Fitting new monobloc taps' (Chapter 10). The only difference you might see is that the basin tap is often connected to the waste by a metal rod. This is the push-button waste mechanism and you can connect and disconnect it by loosening the screw where this rod meets the bar coming out of the waste itself.*

If you've bought a tap with a push-button waste the odds are

you'll want to change the waste as well. We go over this later in 'Fitting a new washbasin'.

Removing an old toilet

The three main types of toilet you might find yourself removing are:

Close-coupled toilet.

Low-level toilet.

High-level toilet.

Height and age are the only real differences between high- and low-level toilets, and these only differ from close-coupled in that the cistern rests on a pair of brackets, rather than the toilet itself, and is connected to the pan via a flush pipe instead of emptying directly into the toilet. So whilst their looks may differ, the steps to removing them are very similar.

Tools and materials

- Pozi and flat-headed screwdrivers
- Sponge or water vacuum
- Adjustable spanner
- Pump pliers
- Utility knife
- Hammer, leather gloves and protective glasses if toilet is cemented in place

1 Turn off the water to the toilet. Ideally you'll be able to do this at the isolation valve close to the toilet. On old toilets this is usually at the side of the cistern, as shown. On newer toilets it's just underneath the cistern. If you can't find an isolation valve turn off all the cold water to the house and then flush the toilet and check it doesn't refill.

2 If you have a push-button mechanism on top of the cistern be very careful when removing the cistern lid as it's often connected to the rest of the cistern and you'll break the flush if you just pull it off.

If it doesn't come up you either need to unscrew the push button system by turning it anticlockwise, or – using a small screwdriver – remove the buttons and then unscrew the large plastic nut this reveals.

Use your fingers to twist the button anticlockwise.

Sadly there are

umpteen variations on this theme so you might need to contact the manufacturer to figure out how to get the cistern lid up without wrecking everything.

If you have a toilet handle just lift the lid off the cistern and use a sponge or water vacuum cleaner to empty it. You might want to remove the ball at the end of the valve to create a bit more space for this.

3 Undo the nut connecting the water supply to the ball valve. The valve can connect to the cistern either at the side or underneath. If it's connected underneath expect a small amount of water to come out of the base at this stage and have a sponge ready.

4 Disconnect the overflow pipe from the cistern.

TIP *Most new toilets don't have or need an overflow pipe, as if the water level gets too high they just empty into the toilet itself. Check the instructions that came with your new toilet and if this is the case remove the old overflow pipe completely.*

---

5 Unscrew the screws holding the cistern against the wall. These are often rusted away, so expect a challenge! If need be use a set of pump pliers to wrench them out.

6 For a close-coupled toilet undo the wing nuts underneath the cistern that hold it in place.
For low- and high-level toilets undo the nut at the base of the

cistern that connects it to the flush pipe and lift the cistern off the supporting brackets.

7 You can now lift the cistern out, remembering that the siphon invariably contains water that's just dying to pour out and get you, and that the base of a close-coupled cistern is usually covered in rust.

8 Empty the toilet pan of water, either by using a water vacuum or by pushing a plunger or toilet brush to and fro to push most of the water out of the U-bend. You don't have to completely empty the U-bend, you just need to lower the level so it doesn't go all over the place when you lift the toilet out.

9 The toilet will either connect to the waste pipe via a plastic connector, in which case just pull it off, or it'll be cemented in place. If this is the case grab a hammer, don a pair of protective glasses and some leather gloves, and gently tap around the toilet outlet to break the connection. You want to be hitting the toilet, not the pipe, as the toilet's being replaced but the waste pipe is going to be reused.

When you've removed the toilet tidy up the waste pipe, removing all the bits of concrete etc.

⚠ **WARNING**

In older houses the waste pipe itself is clay and can easily be broken, so be careful if you're wielding a hammer around it.

10 The toilet is usually held to the floor by screws, bolts, concrete, or a combination thereof. The screws and bolts are usually covered with little caps and can be either through the base or through the side of the toilet pan.

Remove any caps with a small flat-headed screwdriver and undo the screws. You should now be able to lift the toilet up and out, keeping it level so that no water pours out of the U-bend.

If it still seems stuck to the floor just check it hasn't been sealed in with silicon sealant. If it has, cut through the sealant with a utility knife.

If this doesn't work there's a very good chance that the toilet has been fixed in place with concrete. To remove it in these circumstances grab your hammer, glasses and leather gloves again and gently tap the toilet base until it starts to crack and break away.

If you didn't completely empty the toilet in step 8 you might get a little wet at this stage. Once you've cracked all around the base you can remove the toilet pan.

> ⚠ **WARNING**
>
> Broken ceramic is lethally sharp, so whatever you do, keep your leather gloves on until it's all been tidied away.

Fitting a new toilet

There are three main types of toilet you could opt to buy:

CLOSE-COUPLED, STANDARD HORIZONTAL OUTLET

Pardon the pun, but this is the bog standard toilet these days. The outlet from the toilet comes out horizontally, the cistern sits on the back of the pan and it's invariably a two-button flush.

Things to think about

■ More often than not this will be much slimmer than the toilet you're replacing, so you might have to retile behind the old toilet.

■ If your toilet waste pipe comes up out of the floor you need to

check how far away from the wall it is. If it's too far away you'll have a gap between the back of the toilet cistern and the wall. In this case you'll either need to move the waste pipe (not always possible, as it's often set between floor joists) or build some boxing to fill the gap – don't leave a gap, as this will cause the cistern to leak or break sooner or later. You can buy special pan connectors that help get the toilet back closer to the wall, but there's only so much they can do.

■ If your waste pipe comes out of the wall check the height it comes out at. If it's lower than the outlet from the toilet you shouldn't have too much of a problem, but if it's higher you might have to raise the toilet by sitting it on a base.

BACK-TO-THE-WALL

These look good because they hide all the waste pipework. They're also easier to clean. Most come with a horizontal outlet, so the issues are as above plus a few more:

■ To connect it to the waste pipe opt for a 'flexi-pan connector', as these can be extended to allow you

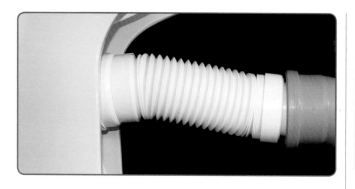

to connect the toilet and waste together, test, check, and then push the toilet back against the wall after you're sure everything is dry and working correctly.

■ A back-to-the-wall toilet will only work if the waste pipe emerges into the room directly behind the toilet. If you have an internal waste pipe that runs across the bathroom to the toilet position

you can't fit a back-to-the-wall toilet unless you hide the waste pipe inside a vanity unit – see below.

■ Many of these come from overseas, where they assume the cold-water supply is coming out of the wall. Here in the UK the supply invariably comes up from the floor. This isn't a problem, but you might find it easier to install if you channel the water supply into the wall first.

BACK-TO-THE-WALL WITH CONCEALED CISTERN

In this scenario the toilet is actually built into a vanity unit. They look great and, because the pipework is hidden, they can be easier to install. On the downside they can be a right pain to service unless you plan for this eventuality when you install it. Other things to consider are:

■ If your waste pipe comes up out of the floor, check the vanity unit depth is sufficient to hide it. If not you might want to build the vanity units a little away from the wall and use wider worktops – or buy a different vanity unit.

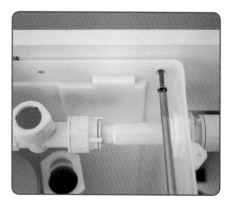

■ Concealed cisterns can leak for years without anyone being any the wiser. One of the main reasons for them leaking is the flush mechanism causing the cistern to lift until it eventually falls off its support hinges. To avoid this, put the cistern on its hinges and then use two screws to keep the cistern in place.

■ The ball valve will eventually fail and need replacing. Make sure you've installed the toilet in such a way that you can turn the water off and replace this ball valve without dismantling the entire bathroom. The same goes for the toilet siphon.

In terms of installation most toilets are the same, so we'll just look at the basic toilet with a push-button flush.

Tools and materials
■ Adjustable spanner
■ Large set of pump pliers
■ Masking tape
■ Pencil or marker pen
■ Drill and tile drill bit
■ Flexi tap connector
■ Pan connector
■ Isolation valve

1 Most toilet cisterns come with the siphon and the ball valve already fitted. If this is the case just check that everything is tight and jump to step 8.

2 There are hundreds of different siphons out there these days, but they'll all have a wide threaded base on to which you push a rubber washer.

3 Check that the inside of the cistern is clean and free of debris then put the siphon through the large hole in the middle so that the washer is sitting inside, on the base of the cistern. If your new toilet uses a flush handle rather than a button, make sure the metal hook attacked to the siphon is on the same side as the hole for the toilet handle. If the toilet is button-operated then it generally doesn't matter how the siphon is arranged inside the cistern.

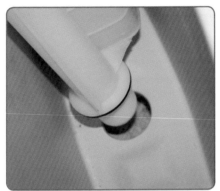

4 Now, things can change here. Your kit will either have a metal plate with two bolts or you'll notice that the cistern has two small holes in the base. If it's the latter, fit the long bolts provided by putting the plastic or metal washer, then the pyramidal rubber washer – pointy end down – on to the bolt before pushing it through the base of the cistern.

Now fit the second washer and tighten everything up with the nut provided. At the end you should have two threaded bolts

sticking out of the base of your cistern, all kept dry by that now squashed, pyramid-shaped rubber washer inside the cistern.

If there are no more free holes in your cistern you should have a metal

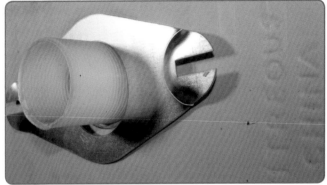

plate supplied with your kit. Either this plate is flat and the cistern has indentations, or the cistern is flat and the plate is shaped. In the latter scenario make sure the plate's the right way up – the ends should be farther away from the cistern base, not closer to it.

Push it over the threaded base of the siphon, check it's aligned properly with the cistern and then tighten the nut at the base of the siphon, using either pump pliers or the plastic spanner that often comes with the kit.

Once tight, slip the bolts into the brackets at each end of the plate.

Note that the base of the siphon usually uses a plastic nut and it'll break if over-tightened, so go carefully.

5 Fit the 'doughnut' seal to the base of the siphon. These come in a variety of shapes, so don't worry if yours isn't exactly like the one shown. However, the wider, thinner, seals are best fitted to the toilet base first rather than the cistern.

6 The ball valve will usually be a short-armed affair and is usually fitted to the base of the cistern.

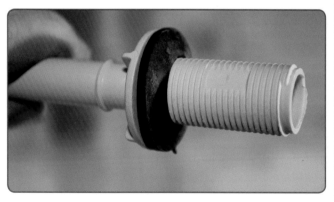

To fit it, push the washer on to the threaded base and push the base through the appropriately sized hole in the cistern. Note that you might find an odd-looking bit of plastic with your ball valve. This is a flow restrictor, designed to help the valve work in areas where the water pressure is high (over 3bar).

If your water pressure is high you push this restrictor into the base of the ball valve.

TIP *Some cisterns come with two holes in the base, both the right size for the float valve. Once upon a time one was used for the float valve and the other for the overflow. These days most toilets overflow into the toilet bowl, so there's no need for an overflow pipe. In this case the toilet kit should come with a blanking disc to seal off the spare hole. Choose the hole that best suits the position of your cold-water pipework to fit the ball valve in and blank off the other by fitting the washer to the blanking plug, pushing the plug through the hole and then fitting the back nut and tightening everything with an adjustable spanner.*

7 Fit the nut to the base of the float valve and tighten it up with an adjustable spanner. You'll need to hold the valve inside the cistern to make sure that it doesn't twist, and you need to check that the float itself can rise and fall freely within the cistern. Note that the nut often has a ridge to ensure that the valve is centred in the hole when tightened. This ridge needs to be facing the cistern base for this to happen.

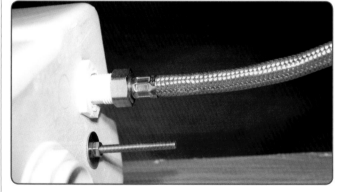

8 It's much easier to fit the toilet to the water supply via a flexible tap connector. You need to fit this to the base of the float valve at this stage.

9 Lift the cistern and carefully place it on the toilet pan so that the bolts on the cistern go through the holes in the toilet pan and the cistern comes to rest on the doughnut seal. The toilet can be a

bit unsteady at this stage so you might want a hand steadying it. Alternatively place it close to a wall so it can't fall over backwards.

10 Tighten the cistern to the toilet pan by putting first a rubber washer, then a metal washer and then a butterfly nut on to the bolts. Tighten each bolt a bit at a time so that the cistern is drawn down level and the doughnut seal is squashed evenly. As usual, there's no need to overdo it; just get everything hand-tight.

11 How the waste pipe emerges into the room dictates what 'pan connector' to use. The general options are:

a **Straight pan connector** – if the waste comes out the wall.
b **Offset pan connector** – if the toilet outlet is slightly higher than or to one side of the waste pipe.
c **Straight extension piece** – if the pipe coming out of the wall is too short.
d **Angled pan connector** – if the waste comes out of the floor.
e **Tight angled pan connector** – if the waste pipe comes out of the floor but you need to move the toilet back farther.
f **Angled and straight flexi pan connectors** – for when all else fails.

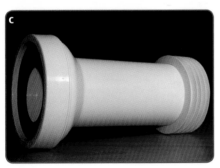

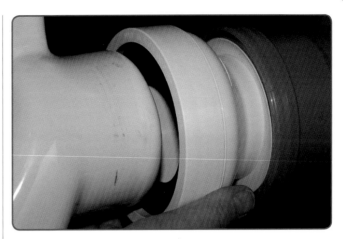

Select the pan connector that works best for your situation and use some silicon grease to lubricate the seals.

12 Fit the pan connector into the waste hole and push the toilet back on to it.

13 Make sure the floor beneath it is even and clear of any debris. Check it sits flat and doesn't rock. If it does you need to make the floor area flatter. If this isn't possible try using slivers of plywood to support the base until you have it steady and level. It's very important that the toilet base is fully supported.

Ensure the cistern is back against the wall. If you can't get it right back you can either build the wall out a little or use short lengths of copper tube to sheath the securing screws and act as spacers. The distance between the cistern and wall will usually determine which option is best.

14 Use a pencil or marker pen to mark the holes for the cistern. Note that some modern cisterns don't have any securing holes. In this case you're expected to fix the cistern to the wall using sealant. If this is the case apply the sealant right at the end of the installation and just use two big blobs behind the cistern – don't overdo it or you'll struggle to get the cistern off in the future!

When you come to drill your holes make sure you drill at a slightly downward angle. If you don't you'll really struggle to fit the screws.

15 Mark the holes for the toilet base. There are usually two approaches to this:

There are holes in the base of the toilet

If you're putting the toilet on to a wooden floor then there's no problem. If the floor is tiled you'll need to mark and drill the holes first. Try to put a pencil or marker pen through the holes to mark the floor directly. If this doesn't work:

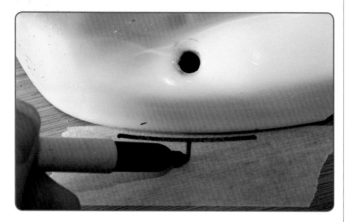

Put masking tape around the base of the toilet and mark the toilet edge and where the holes are.

Lift the toilet and put masking tape underneath, roughly where you think the holes are going to be.

If a pencil or pen won't go through the hole use the screw you're going to fit. Ideally this will pierce your masking tape and leave a mark. When using the screw to make your mark remember to push it in at the angle you'll later screw it down at, *ie* at an angle that will allow you to use your screwdriver.

If it doesn't, mark the tape rub a pencil over the end of the screw and then push the screw in again. This will usually leave a tiny pencil mark on the tape.

If this still doesn't work measure the distance from the hole to the edge of the toilet, then remove the toilet and mark the masking tape accordingly.

There are holes in the side of the toilet pan

In this instance the toilet is usually supplied with little plastic L shapes (usually blue).

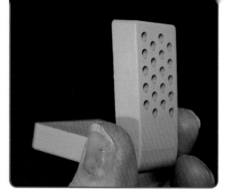

To fit these, put masking tape around the edge of the toilet and run a pencil around the base.

Mark the position of the holes in relation to the base

Remove the toilet and measure the width of the toilet base.

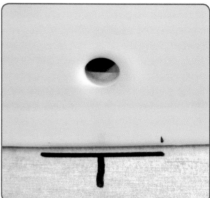

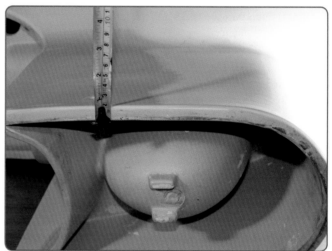

Apply more masking tape under the toilet and mark a little bit more than the width of the base.

Put the blocks into position and mark the hole. If it's a wooden floor just screw the blocks into place, remembering to check that there's no pipework under the floor. If the floor is tiled you need to grab yourself a drill and an appropriate bit.
  Having drilled all your holes fit the appropriate rawlplugs.

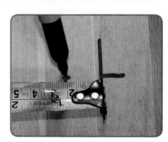

TIP *For some reason toilets come with enormous bolts and screws to bind them to the floor. I can see no logic for this. After all, the toilet is already on the ground – where's it going to go? Alas, it's always best to use what the manufacturer supplied so prepare yourself to drill a great big hole.*

Things to ponder before you do this

■ *Have you fitted underfloor heating? If so, did you fit it under the toilet position? If the answer might be yes then don't screw the toilet to the floor – use silicon sealant instead.*

■ *How long are your screws? If they're going through floorboards are you sure no pipework is underneath them? If in doubt either use shorter screws that won't pierce the floorboards completely, or use silicon sealant instead.*

■ *If you're going through floor tiles use a tile drill and work your way up to the hole size you need, ie start off with a 6mm drill bit, then go to 8mm and finish off with 10mm.*

■ *If the floor tiles are porcelain you'll need several drill bits to get through the tile ... and bring some lunch, as this might take some time.*

■ *Don't use the drill's 'hammer' setting on tiles.*

In days of old they used cement to fix a toilet to the floor. This has gone out of favour these days, mainly because cement has a tendency to heat up and expand, causing the toilet pan to crack – so don't use it. If you really don't fancy screwing your toilet to the floor silicon sealant does a perfectly good job when applied correctly.

16 Now, you could screw everything down at this point, but I prefer to test things out first. So let's connect the toilet to the cold water. The odds are that the original pipework doesn't fit, so the first stage is to cut it back and fit an isolation valve – ensuring the arrow is pointing in the right direction, ie towards the toilet.

If luck is on your side and the original pipework does fit, *still* fit an isolation valve if there isn't one already there.

17 The easiest way to make the connection to the toilet is to use a ½in–15mm flexi tap connector. Remember not to bend or twist the hose too much or you'll restrict water flow. To stop it twisting hold the base of the hose in place with a set of pump pliers whilst you tighten the nut with an adjustable spanner. Alternatively buy the push-fit version of the connector.

18 You can now turn the water back on and check that everything fills properly and all is dry.

19 Check that the water level in the cistern is right – there's usually a line in the cistern to show the correct level, although it's rarely obvious. If the level's not right, adjust the height of the float. Each float is different, so check the manufacturer's instructions, but there's usually a threaded bar joining the float to the top of the

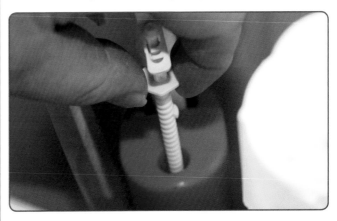

valve. Turn this clockwise and the float height is adjusted up, turn it anti-clockwise to lower the water level.

20 Once you're sure that all is dry, flush the toilet. Again, how you do this varies from siphon to siphon but there's usually an obvious part to either lift or push.

21 If all remains dry after the flush secure the cistern to the wall either with sealant or with the screws you've already drilled holes for. It's best to use plastic washers to stop the screw chipping the ceramic, and don't overdo it or you'll crack the cistern. The screws will rust over time so put plastic caps over them to protect them as much as possible.

22 Do the same for the base. Again, use the washers and caps that usually come with the new toilet.

23 You're now ready to fit the cistern lid and the push-button mechanism. There are umpteen ways of fitting these buttons so you're really going to have to read the

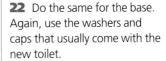

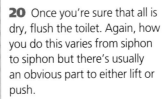

manufacturer's instructions to figure out what to do. However, here are a few tips:

- If, having fitted the push-button mechanism, you notice that the toilet is constantly leaking a small amount of water into the pan either the pegs attached to the buttons are too long, or you've set the siphon too high.
- Some push-button mechanisms are linked to the cistern lid and the lid can no longer be lifted off without breaking the mechanism. Explain this to everyone in your household, because, deep in the DNA of homo sapiens, there's an overwhelming urge to peer into the cistern at some point. Fortunately manufacturers seem to have caught on to this so most new cistern lids can be lifted without causing damage.
- Don't use those little blocks of stuff that turn the flush water funny colours in your new cistern, as most modern siphons will start leaking if you do. If you can't live without coloured flush water opt for the under-the-rim variety.
- If your new toilet uses a handle read 'Replacing a broken toilet handle' in Chapter 12.

24 Now all that remains is the toilet seat. There are some extremely rare toilet seats that need fitting before the toilet's in place, but I appear to be the only person who was ever foolish enough to buy one of these so I'll assume that the seat in your possession needs to be fitted now.

Like push-button mechanisms, each manufacturer seems determined to design a seat that fits slightly differently from every other seat. Fortunately there's only so much you can do with a toilet seat, so they all share some common features which we'll look at now:

On the top of the toilet pan are two holes that accept the seat fixing brackets of whatever mechanism you've been supplied with.

The basic design just has a screw with a washer, something to hide the screw once the seat is fitted, and some sort of mechanism that allows you to adjust the seat to fit the shape of your toilet – usually just an elongated groove in a metal washer.

Having pushed the mechanism into the holes you either tighten them up from the top, or you drop to your knees and tighten the screw tail underneath.

Once the seat is firmly secured in place you cover up the mechanism with either a plastic or metal cap.

Removing a bath

CAST IRON BATHS

- These weigh an absolute ton. To remove them disconnect the waste and water as explained below.
- Now cover the bath with an old blanket or towel.
- Don a pair of ear defenders, leather gloves and protective glasses and grab a large hammer – the bigger the better.
- Prepare everyone for an awful lot of noise and start hammering away.
- Break the bath into portable chunks and remember that the edges can be very sharp.

STEEL BATHS

These can be removed exactly the same way as a standard acrylic bath, as detailed below. They usually don't weigh much but it's best to get a friend involved when removing them from the house.

TIP *At the time of writing scrap metal was worth a fortune, so it's well worth taking your old cast iron or steel bath to the local scrapyard.*

ACRYLIC BATHS

Most modern baths are made of acrylic and fibreglass and are fairly light. However, they're still pretty big so rope in a friend to help you get it out of the house.

TIP *It's nigh on impossible to remove a bath without damaging at least some of the tiles around it. Bear this in mind if you're not thinking of retiling your bathroom afterwards.*

Tools and materials

- Screwdriver set
- 15mm and 22mm pipe-slice
- Adjustable spanner
- Utility knife

1 Remove the bath panel – usually there are just a few screws under little decorative caps that hold the panel in place. The panel

will usually be secured with sealant as well. If this is the case just run a utility knife along the seal to break it. You normally have to pull out the bottom of the panel first, which releases the lip at the top.

2 If there are isolation valves fitted, close them. Otherwise turn off your hot and cold water and drain the bath taps until the water stops running; then open the kitchen taps downstairs and wait until they stop running.

TIP *The cold water to a bath is sometimes from a stored water source. As such, just because all your other cold-water taps have stopped doesn't mean your cold-water bath tap has.*

3 Use an appropriate pipe cutter to cut through the pipework to the taps. This is quicker than removing the taps, but if you'd prefer you can do this instead see 'Replacing washbasin and bath taps' earlier in this chapter.

 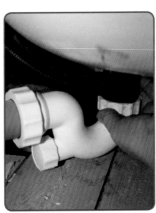

4 Undo the waste trap at the base of the bath – this is usually only hand-tight. As you lower the trap you might get a release of dirty water, so put a sponge underneath.

5 The bath legs are usually held steady by two nuts, one above the fixing bracket, one below the bracket. Use an adjustable spanner to loosen the bottom one, allowing the bath to drop a little. Do this for all the legs.

6 The legs are often held to the floor by little screws. Undo these so the legs are loose.

7 Run a utility knife around the bath to break the sealant line.

8 The bath is often held to the wall by two little brackets that are now, more often than not, behind some tiles. You should have some 'wiggle-room' with the bath, and giving it a shake should give you a clue as to where the brackets are. If you can remove these tiles without breaking them you should consider entering the lottery this weekend. Otherwise break the tiles and unscrew the brackets.

9 The only thing now holding your bath in place is either more sealant, which you can usually overcome by just pulling hard, or the bathroom walls. Many baths are recessed slightly into the walls of the bathroom, partly for extra support, partly because the bath is slightly too long for the room. If this is the case you'll probably have to remove the first few rows of tiles above one side of the bath and lift this end out first to free the bath.

Fitting a new bath

Installing a bath is pretty standard so we're just going to step through the basic acrylic offering. However, before we start there are a few things to think about. Firstly, what kind of bath do you want?

CAST IRON BATHS

This is often the first choice of the overly romantic. Alas, there's a good chance of you putting your back out just getting it into the house, as they can weigh an enormous amount – you'll need at least three people to get one up a flight of stairs. Once in you might have to strengthen the floor a little to support it, and once in use you'll find that they can be more than a little cold to start off with and are easily chipped and damaged.

STEEL BATHS

These are often the cheapest option. They don't tend to flex as much as acrylic baths and are light enough to readily fit, but they're prone to chipping and can be cold when you first get in.

FREE-STANDING BATHS

As with cast iron baths these are for people who read too much Mills & Boon and have an ominous penchant for scented candles. Yes, a nice free-standing slipper bath might look like something out of *Cinderella* but you need a bloody big bathroom for the

'look' to really work, and since all the plumbing is on show they can be a right palaver to fit.

Modern ones are often made out of really thick acrylic that reduces the weight whilst still looking good. All invariably use chrome or stainless steel waste fittings, which never fit standard waste pipework properly and need to disappear under the floor or into a wall pretty pronto – so if you have concrete floors you're going to struggle to fit a free-standing bath.

The water pipework is usually sleeved in a chrome or steel sheath. Sometimes these have been designed by someone with an ounce of sense and are telescopic and wide enough to accept an isolation valve; but often they've been designed by someone who knows as much about plumbing as I do about astrophysics and are a right pain to install as a result.

SHOWER BATHS
These are an ideal compromise for those of us who prefer a shower most days but don't want to lose the luxury of an occasional bath. They're shaped so that there's a larger showering area, but the overall size of the bath is largely unaffected. As such they still fit into most bathrooms.

ACRYLIC BATHS
If you bought a cheap bathroom suite this is where the manufacturer has usually made most of the cost savings. I assume there's a minimum weight that a bath must be able to support, and having a looked at quite a few baths I think it approximates to that of an obese dormouse.

Some baths are ludicrously thin (5mm) and have such spindly little aluminium legs that they look like they'd struggle to hold themselves up, let alone the person sitting in them. Personally I'd try to get a bath that's a minimum of 8mm thick and has legs that look like they'd have a fair chance of supporting me after a weekend at a pie festival.

Other things you'll need to consider, regardless of the bath type are:

■ A standard bath is 1,700mm long by 760mm wide. If the gap it's to fit in is smaller than this you'll either have to channel out the wall to accept it or buy a slightly smaller bath. You don't want to channel out more than a few centimetres at either end. If you need more than this buy a smaller bath.

■ The bath taps need some space behind them to be easily operated. Bear this in mind when channelling out the wall – don't channel out more than a centimetre for the end where the taps are going. If this can't be avoided think about buying a bath with the taps in the middle.

■ If the gap for your bath is bigger than 1,700mm you'll either need to build a shelf at the end of the bath away from the taps and shower, or buy a bigger bath – 1,800mm is readily available.

■ If you're fitting a whirlpool or spa bath you'll need to fit a fused spur, and you'll need ready access to the electric pump. Electricity and water don't mix well at the best of times so the fused spur is best located outside the bathroom and should be fitted by a qualified electrician.

■ If you've removed an old bath it's not unusual for the floorboards below to be water-damaged. A full bath weighs a fair amount, so always check and replace the floorboards if need be.

■ It's possible to buy big 'double' baths. When full these can weigh an enormous amount, so if in doubt strengthen the area where the legs will stand or risk bathing in the living room one day.

We're going to start off by assuming that you've checked your bath is the right size for your bathroom, an obvious but often overlooked starting point.

Tools and materials
■ 22mm and 15mm isolation valves
■ Pump pliers
■ Adjustable spanner
■ Level
■ Tape measure
■ Masking tape
■ Pencil/marker pen
■ 2in x 1in timber (approx 3.5m)
■ Saw
■ Drill (with hammer only action if you're going to channel out the wall) or a hammer and bolster
■ 2mm wood drill bit
■ 8mm masonry drill bit
■ 8mm rawlplugs
■ 50mm size 10 screws
■ Screwdriver/impact driver
■ Hole-cutting set
■ ¾in box spanner
■ Tubes of clear and white sealant
■ Utility knife

1 If you're just replacing a bath then the hot and cold and waste pipework is probably in pretty much the right place already. It's nigh on impossible to reuse the entire pipework, so having ensured the hot and cold pipework's been drained cut it back to about 8cm off the floor and fit isolation valves – checking that the arrow is pointing in the right direction.

If your bath is fed by stored water you might want to use full-bore isolation valves so as not to restrict the flow too much.

If isolation valves are already fitted, close them and then undo the nut at the top of the valve, holding the valve steady whilst you do so, and remove the old pipework.

TIP *In the UK old pipework is often imperial size rather than 22mm. It's close to impossible to tell the difference by looking at them, but the imperial ¾in pipe is slightly narrower. As such you'll notice that 22mm olives and soldered fittings seem suspiciously loose. Push-fit copes with the difference fine, but you'll have to fit ¾in olives to compression fittings and get ¾in–22mm adapters for soldered fittings.*

For 15mm pipework the difference is so negligible you can get by without worrying about it – you'll just notice that it's a bit of a tighter fit with imperial pipe.

Now put a length of pipe about 20cm long on to your isolation valves and, ensuring the olive is in place, tighten it into position. This is too long for what we need but too long is better than too short. The only reason we're doing all this is because it's easier to tighten and test these valves whilst there's no bath in the way.

Now close the isolation valves, turn your hot and cold water back on and check for leaks.

For the waste pipework, just check that you'll be able to get at it once the bath's in place so you can make any minor adjustments.

Now jump to step 3.

TIP *If you're reversing the bath – ie the taps are going to be at the opposite end to where they were – then you're best off putting an elbow on the pipework before fitting the isolation valves. Also, remember that you have to cross the pipework over at some*

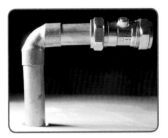

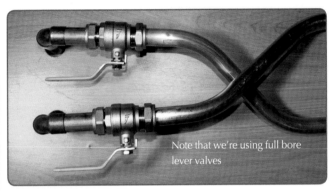

Note that we're using full bore lever valves

point, otherwise you'll end up with hot water coming out of the cold and visa versa.

Another thing you have to bear in mind if you're running pipework under the bath is the location of the bath feet.

2 If the bath is going into a new position you'll need to get the pipework into place first. Generally you'll want it emerging above the floorboards about 8–10cm from the wall, with the hot and cold about 10cm from the centre line of your new bath, *ie* about 20cm apart.

What diameter pipe you use depends on what type of hot- and cold-water system you have – see Chapter 2. If the bath uses stored hot and cold both pipes need to be 22mm in diameter. If the bath uses stored hot and mains cold then the cold is 15mm and the hot 22mm. If the bath uses high-pressure hot and cold, *ie* you have a combi boiler or an unvented hot-water cylinder, both pipes can be 15mm.

TIP *The hot-water tap should always be on the left, the cold-water tap on the right. The easiest way to remember this is to think of the UK weather, 'right cold'.*

If you're running hot and cold pipework close to each other, the hot should always be on the top. Do it the other way around and your cold water ends up warm when it comes out of the tap. If you can't avoid having the hot under the cold insulate one or other of the pipes.

Once the pipework's in place, fit isolation valves, close them, turn your water back on and check for leaks before replacing any floorboards.

The waste pipework needs to be in 40mm pipe and shouldn't run more than 3m before reaching the soil stack. If you're going to exceed this use 50mm pipe from the soil stack and reduce it to 40mm just before the bath, or extend the 110mm soil stack pipework itself.

Remember to avoid using push-fit waste pipework under the floor – go for solvent weld.

You want to aim for a drop in the pipework of 18–90mm per metre length. If the drop is greater than this you risk pulling the water seal out of the bath trap every time you empty the bath, which lets noxious smells into the bathroom. If the gradient is less than this the bath isn't going to drain well and the pipework will have a greater tendency to block up.

There are a number of ways of connecting to the soil stack itself:

Either put a rubber reducing seal into the end of a 110mm socket.

Use a hole-cutter to cut out one of the bosses already on the stack and fit a 40mm boss.

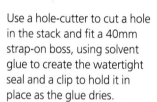

Use a hole-cutter to cut a hole in the stack and fit a 40mm strap-on boss, using solvent glue to create the watertight seal and a clip to hold it in place as the glue dries.

3 The first step is to decide how high you want your bath to be. The bath height is really a compromise between ease of getting into it versus getting a bath trap in under the bath and getting a decent drop on the waste pipework.

If you purchased a plastic bath panel with the bath this effectively sets the maximum height for you, whilst the waste pipework will determine the minimum. If you bought a two-piece panel, where the height can be adjusted, aim for a height of about 500–540mm – higher if your waste is already above ground level, lower if it's coming up out of the floor under the bath.

If the waste pipework is coming out of a solid wall and is therefore set and can't be lowered you might want to fit the bath legs first and set the height according to this waste pipe.

In this example we're going to assume you have a fixed-height plastic panel and we're going to set the height first based on this panel. However, feel free to jump to step 14 and fit the bath legs and use the waste pipe to determine the minimum height before coming back here.

Measure the height of the bath panel, from the edge that tucks into the top of the bath to the base, and jot this down.

4 You're trying to set the height of the bath so that you don't have to cut the bath panel down if it can be avoided. In an ideal world the floor will be perfectly flat, but there's a good chance it isn't and what you don't want to end up with is a huge gap at the bottom of the panel at one end. To avoid this you want to set your height at the end of the bath where the floor is lowest, so grab a level and check the pitch of the floor.

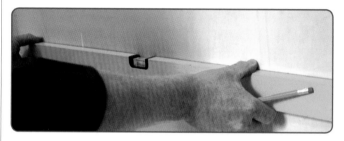

5 Having determined which end is lowest, use the height you jotted down earlier. Mark the wall and use a level to draw a horizontal line around the wall where the bath will fit.

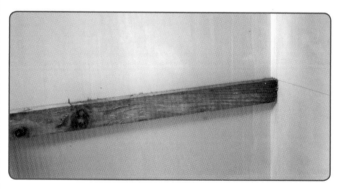

6 Cut some perfectly flat 2in x 1in timber so it just fits into the long gap, with the top of the timber against the line we have made.

7 If you have a solid wall you can screw the timber to the wall where you like, so jump to step 8.

If you have a stud wall you'll need to screw this timber to the wall where the battens are. Fortunately the wall is rarely plastered underneath the bath so you should be able to see where the plasterboard's been screwed or nailed to the battens, and can use these marks.

If you can't see where the batons are it's best to buy a stud detector. These are handy to have for most DIY jobs as they find not just the wooden battens (or studs) but also hidden metal, *eg* water pipes and electrical cabling. However, keep in mind that stud detectors aren't absolutely reliable. Just because a detector says there's no cable or pipework hidden in a wall doesn't mean there isn't – remember, it won't detect plastic pipe. So use them as a guide only and proceed with care.

Having found the studs (battens) mark them on the wall and put your timber in place.

8 Mark the timber with an arrow pointing up. This will stop you getting confused as to which way round the wood is later on.

9 Mark at least three holes on the wood, either where the studs are or, for a solid wall, about 10cm from each end and roughly in the middle.

10 Use a 2mm bit to drill through your marks – if you don't the timber has a habit of cracking as you screw through it.

11 Set the timber in place with the top against the line you drew.

12 To get the wooden supports perfectly level, it's easiest to fully drill one hole initially and screw the wood to the wall. Then use the level to get the wood perfectly horizontal again before marking and drilling the other holes. You can move the timber out of the way while you drill them.

13 If you're not having to channel out the walls to fit the bath you can now just repeat steps 6–12 for the other walls the bath's going to be up against. Remember, though, that the timbers want to be about 40mm shorter than the width of the bath so that you can get the bath panel in OK.

The easiest way to ensure that the end battens are level is to screw them in at the end nearest to the already installed batten and then use the level across both pieces of wood.

If you're less than convinced about the strength of your wall feel free to add some vertical supports too.

If you're going to channel out the wall, now's the time to do it. The line you've drawn on the wall is where the bottom lip of the bath will rest, so you need to channel out from here and then up for the thickness of the bath lip – usually about 40mm.

Most drills these days have a 'hammer only' setting and you can buy channelling tools for them. If this all sounds too much, get a

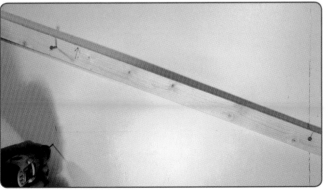

bolster and a hammer and start hitting your wall – you're usually only removing the plaster and the rendering and possibly a bit of breeze block, so it's all fairly soft.

Mark out the edges of the channel first and then remove the plaster between these marks.

TIP *If the bath is a really tight fit you're either going to have to channel out quite a height on one side of the bath and fit it at an angle, or make a longer channel in the wall so you can slide the bath into place.*

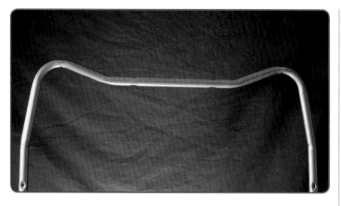

14 The bath usually arrives with the leg supports flat-packed against it. Unwrap the bath now and remove the supports.

There are two common approaches to bath legs. They're either spindly little things that fit into plastic feet or more substantial metal bars on to which the bath feet fit. Personally I prefer the latter, but the former seem to be more common so we'll follow the directions for these.

15 Lay the bath into your timber frame and check all is level across the length of the bath and the ends, and also diagonally across the bath. If it doesn't seem level check your battens and make any adjustments necessary.

16 You're now ready to 'dress' the bath, *ie* fit all the bits to it. The bottom of the bath is usually reinforced with wood covered by fibreglass. The two leg supports need to be on this wood. One needs to be just after the waste outlet but not so close that it'll interfere with the waste itself; the other needs to be fairly close to the end of the bath. You'll be left with one little bracket that supports the central leg. In the photograph I've shown the rough positions in black marker pen. Mark your holes with a pencil or marker pen.

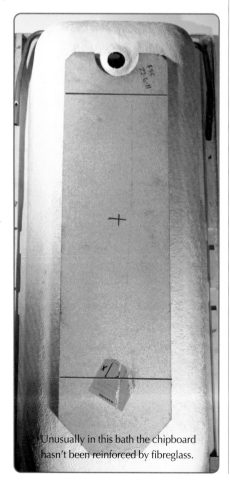

Unusually in this bath the chipboard hasn't been reinforced by fibreglass.

17 Screwing directly through fibreglass is difficult so use a 2mm drill bit to make a pilot hole first. The hole only needs to be about 2mm deep so put a stop on the drill to prevent you going right through and into the bath. Alternatively use my preferred approach and just be careful.

18 Fit the plastic covers on to the leg supports and screw the supports on to the base of the bath. Make sure you use

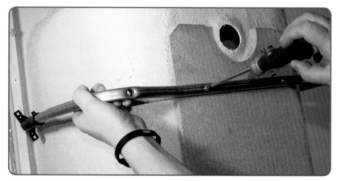

the right screws – people screwing the legs on to a bath have been known to screw right through into the bath itself. If in any doubt opt for the smallest screws provided. *Do not use the screws that held the legs to the bath during transit.*

19 Make sure the legs are perpendicular to the bath and then screw the plastic supports into the wooden edge of the bath.

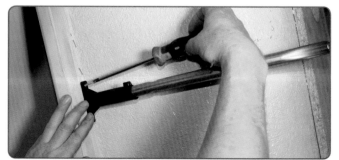

20 Screw through the plastic supports into the side of the leg to secure it in place.

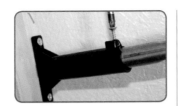

21 The legs we're using here come with a central support bracket into which the shortest of the supplied legs will fit. Fit this leg loosely for now. If your bath comes with two metal bars to which the legs are attached there usually isn't a central leg.

22 Fit the legs to the supports by running one nut down the threaded leg, pushing the leg into the support and then adding the second nut to tighten it up – just hand-tight for now.

You know the bath height already so measure the distance from the base of the leg to the bottom lip of the bath. You want to set the legs just a little lower than this height.

Be aware that the centre leg is often shorter than the rest. When you first fit the bath you don't want this leg taking any weight, so set it lower than needed and leave it fairly loose.

TIP *Work out the height for one leg and use this as a template for the others.*

23 Before you fit the taps and waste you need to remove the protective film on the bath. This is almost impossible to see and I once spent half an hour trying to remove a film that wasn't there. However it usually is, so start picking around the waste hole until you can start to peel the film back. You don't want to remove it all, just peel it back from the waste, overflow and tap holes and from around the edges.

24 Sometimes the bath doesn't come with pre-drilled holes for the taps. This gives you far more flexibility but does create the anxiety of having to cut your own holes. If you already have a set of holes jump to step 25.

To cut your own holes

■ Cover the area where the taps are going with masking tape.

■ Measure and mark the centre of the bath.

■ Separate bath taps are largely a thing of the past so your tap is probably a single block of brass with two tails – one for hot water, the other for cold. If this is the case you need to measure the distance between the two tails. Having done that measure them again, as a mistake now would ruin your bath. Rather than measure the centre of a hole it's usually easier to measure the left of one hole to the left of the next, or else right to right.

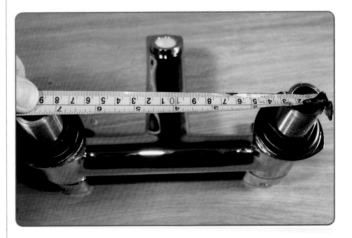

■ If you do have two separate taps set them about 18cm apart.

■ Mark this distance on the bath, using your centre line. Check it again. (The old adage 'measure twice, cut once' is remarkably apt.)

When deciding how far forward or back your taps need to be, consider the following:

- If you're channelling out the wall how much of the end of the bath is still going to be available?
- The tiles will encroach about 1.5cm over the back of the bath.
- There's a wooden edge under the bath – check you're not trying to put the taps through, or too close, to this.

- Check the length of the tap spout and make sure it'll overhang the bath.

Use a hole-cutter that's just big enough to accept the tap tails and carefully cut your holes.

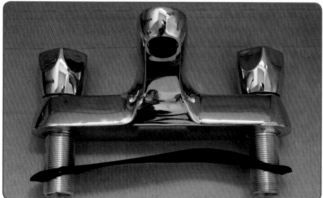

25 Some taps use a rubber or foam base to create a watertight seal. Others use a rubber washer under the base of the tap. In either case these

create a seal at the top of the bath. With the seals in place push the tap through the holes in the bath.

26 Fit the nuts to the tails and hand-tighten. Then use a ¾in box spanner or an adjustable spanner to get them really tight.

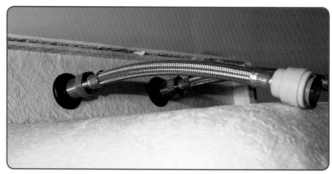

27 Hand-tighten two flexible tap connectors to the tap tails. Push-fit connectors are more expensive but much easier to work with. Once hand-tight give them an extra three-quarter turn with an adjustable spanner.

Often the cold water is in 15mm whilst the hot is in 22mm, in which case you'll need one 15mm–¾in flexi tap connector and one 22mm–¾in flexi tap connector.

Though flexible tap connectors make fitting the bath much easier, they do restrict the flow a bit. If you have high-pressure hot and cold water this shouldn't be an issue, but for baths fed from storage tanks you might want to buy connectors with wider diameters or use plastic pipework and fittings. This is particularly important if you're fitting a large bath that will take an age to fill at the best of times.

28 Fit the bath waste. These vary enormously but a common type has the overflow and the waste linked together and turning the overflow cover pops-up the bath waste. Don't worry if you don't have one of these as the principles remain the same: a washer under the bath makes the seal, the base of the waste has a connector to link it to the overflow, and its base is threaded to accept a bath trap.

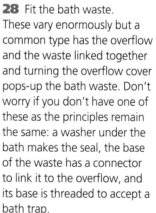

29 Fit the washer to the base of the bath waste – note that it's invariably shaped to fit into the top of this base.

30 Hold the waste in place under the bath as you fit the

top of it into position in the bath. Now tighten the two together using the thick screw supplied. I find an old chisel is the best tool for tightening this nut.

31 Fit the overflow by first removing the chrome cover.

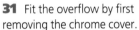

32 The mechanism revealed needs to be dismantled. First remove the central screw, then pull the front of the overflow away from the rear mechanism.

33 Put the base of the overflow, with its washer, in place. Put the front of the overflow on the base and screw it into position – hand-tight should be enough to create a watertight seal. Finally, push the chrome cover into place, making sure the grub screw is at the bottom, and tighten it up.

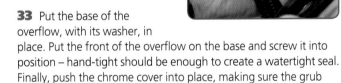

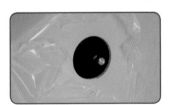

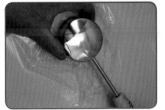

34 Some baths come with metal arms to help you get in and out, and if yours has these now's the time to fit them.

With everything fitted you can now put the bath temporarily in place, checking that it's resting on the timber supports rather than its feet, and that all is level.

If you have a pop-up waste you might want to adjust the height of the plug and check that the mechanism works at this stage.

35 You should have more pipework under the bath than you need. So lie down, hold the ends of the tap connectors in place and mark the pipework. You can see how deep into the connectors the pipework will fit by just looking at the connector but as a guide it's 30mm for 15mm fittings and 38mm for 22mm fittings. Since you have flexible fittings it's best to err on the side of caution and cut the pipework a little longer than you think you'll need.

Try to keep bends in the connectors to a minimum, as tight bends will restrict water flow.

If you're fitting taps to the middle of the bath you're unlikely to get access to them once the bath is in place, so for these it's best to either use push-fit pipework and connect long lengths to the taps before fitting the bath, or else buy very long flexi tap connectors. *Don't* use washing machine hoses, tempting though this may be.

36 Put the ends of your tap connectors over your cut pipework and check the pipework goes fully into the fitting and that the connectors aren't bent too much. We're using push-fit in this example but it you opted for compression fittings don't tighten them yet as you're about to take the bath out again.

37 You now need to fit a U-bend – more commonly called a 'trap'– to the bath. There are two main bath ones you'll come across – the 'shallow P-trap' and the 'standard bath P-trap'.

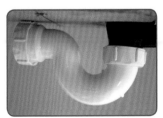

The former is often the easiest to fit, but according to the building regs it should only be fitted to a bath if the waste pipework terminates in an open gully. If your bath waste pipework goes directly into a soil stack – and most do – you should fit a trap that's at least 75mm deep. The problem is that these traps usually require you to cut out the floor under the bath in order to fit them, and Sod's Law often kicks in at this point and you discover a joist directly under the place where the trap is supposed to go.

A way of getting around this is to fit a 'self-sealing' trap. These work by using a rubber membrane rather than water to create the seal. Because they don't need a depth of water they can be very shallow – ideal where your bath is low and over a concrete floor. Sadly they don't seem to make ones that connect directly to the bath waste – although they do fit directly on to a basin waste – so you need to fit it on to the waste pipework – which normally requires an adapter.

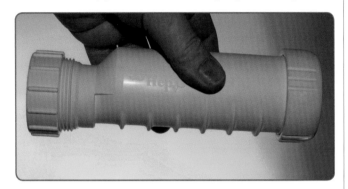

The reason why a traditional trap needs to be so deep is to stop the water being sucked out of it when the bath empties. If this happens you end up releasing the smell of the soil stack into your bathroom. This can also happen if you have a bath that you don't use very often. In this case the water in the trap evaporates and let's smells in. If this is something you have trouble with replace the trap with a 'self-sealing' type.

38 Having selected your trap – and cut any holes in the floor that you need – hand-tighten it to the bath waste and connect it to the waste pipework. Remember that you

can change the angle of the trap to accommodate your pipework, so play around to find the angle that's going to work best.

Once you're sure the waste is in right, undo the trap from the base of the bath, disconnect the tap connectors and lift the bath out.

39 Everything is now ready for you to install the bath. First off, run a thick line of sealant just above and on top of your timber frame.

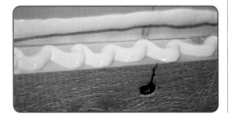

40 Put the bath carefully into position, push it back against the wall and down against the sealant. Check that it's level.

TIP *Some of you might be wondering why we're not using the little brackets that came with the bath to hold it back against the wall. Well, the sealant is doing this for you and creating a waterproof seal at the same time, so you don't really need these brackets. If you really want to you can fit them just before step 40, but you'll have to notch out the timber frame first to accept them.*

41 Adjust the legs so that they're securely on the floor. It's possible to screw down the legs closest to you and this will keep the bath in position whilst the sealant sets.

42 Connect the flexi tap connectors to the hot- and cold-water pipework. If you're using compression types hold the hexagonal base to the fitting steady to stop the hose twisting.

43 Connect the trap to the bath waste – it only needs to be hand-tight but give it an extra tweak with a pair of pump pliers if you have any doubts.

44 Close the taps then open the isolation valves and test all is dry. If so, open the taps and let the water flow out of the plughole. If all is still dry put the plug in place and fill the bath. When the water is about 30mm deep open the plug and again check for leaks.

If all is dry fill the bath with about 400mm of water. This gives the bath some weight and ensures it settles into place.

45 Get under the bath and start tightening all the legs, apart from the centre one. Once you've done that, check the bath is still level. If it isn't adjust the legs to get it level, making sure the frame is still taking the vast majority of the weight. When all is fine set the central leg.

46 Remove any excess sealant from around the bath using a sealant profile or your finger and some dry tissue. Also check for

gaps in the sealant. If there are any, fill them in and smooth the sealant back as described earlier.

47 Leave the bath overnight. This lets you check for leaks as well as allowing the sealant to start setting. If all is well, fit the bath panel.

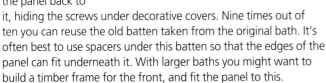

To get a nice fit for the panel lay a piece of timber along the floor so you can screw the panel back to it, hiding the screws under decorative covers. Nine times out of ten you can reuse the old batten taken from the original bath. It's often best to use spacers under this batten so that the edges of the panel can fit underneath it. With larger baths you might want to build a timber frame for the front, and fit the panel to this.

If you have to cut a plastic panel, carefully use a utility knife to score the surface – ideally by running the blade down the side of a level or other straight edge – and then bend the plastic to break it.

TIP *Whilst the bath panel is a decorative item it's also the service hatch for the bath itself. As such you need to make sure it can be removed without too much hassle.*

Removing a washbasin

Tools and materials
- 15mm pipe-slice
- Adjustable spanner
- Set of screwdrivers
- Utility knife

1 Turn off the water and check it's off by opening the basin taps. Ideally you'll be able to do this by closing the isolation valves. If these haven't been fitted turn off the hot and cold to the house and open up the kitchen taps to drain the water away.

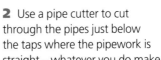

2 Use a pipe cutter to cut through the pipes just below the taps where the pipework is straight – whatever you do make sure you're cutting the pipework above the isolation valves!

3 Disconnect the waste pipework, ideally by undoing the trap. Sometimes you need to move the pedestal aside before this is possible, so you might want to jump to steps 4 and 5 and then come back to this point.

4 Look under the basin. It is usually bolted or screwed to the wall. If it is, use a spanner or screwdriver to release it.

5 Use a utility knife to cut through the sealant at the top of the washbasin.

6 At this stage you should be able to carefully lift the basin out, making sure the pedestal – if you have one – doesn't fall over.

7 The pedestal is often free-standing and can now just be lifted away. However, sometimes they're screwed or stuck to the floor with sealant.

Fitting a new washbasin

There's quite a range of washbasins you might consider fitting:

FREE-STANDING WASHBASIN

These look great in the showroom but the design often makes it difficult to get the water and waste pipework out without sullying the 'look'. To keep them looking good you often need to bury all the pipework in the wall, which is a time-consuming business and not always feasible. Some of these come on little stands where all the weight is taken by the stand itself, which is ideal if your bathroom has stud walls.

NOTE
A stud wall is one where the wall consists of a timber frame, filled with insulation – usually – and covered by plasterboard. Whilst they're easy to build they're not strong and you shouldn't hang anything off them without first strengthening the wall by putting either a batten or a sheet of 19mm plywood between the studs.

INSET WASHBASIN AND VANITY UNIT

These give you lots of bathroom storage space and keep all the pipework hidden. Since the basin is supported by the unit they're ideal up against a stud wall – assuming the unit is floor-standing.

SEMI-RECESSED WASHBASIN

Very like the vanity unit but much thinner. They still provide some storage and hide the pipework but the unit is less intrusive. They're ideal in a smaller bathroom where an inset basin takes up far too much space.

WASHBASIN WITH NO PEDESTAL

These work well in very small cloakrooms but you need to bury the pipework in the wall if they're going to look any good. The weight of the basin is taken by the wall so you either need a solid wall or will have to strengthen a stud wall by putting a batten or plywood into it. Since the waste is on show you might want to consider a chrome trap and waste, with all the hassle that brings.

WASHBASIN WITH SEMI-PEDESTAL

These are a bit of a compromise: the wall still takes the weight but the pipework and waste is hidden behind the pedestal, giving a cleaner look.

WASHBASIN AND PEDESTAL

This is the most common washbasin you'll come across. The pedestal is primarily there to hide the pipework but it also supports some of the weight of the basin. Having said that, the vast

majority of the weight is taken by the wall, so you should consider strengthening any stud wall the basin is up against.

We're going to step through the fitting of a washbasin with a pedestal, but the plumbing side is pretty much the same for all of them; the waste needs to run through 32mm pipework and the hot and cold are in 15mm pipe.

Tools and materials

- ■ 15mm pipe-slice
- ■ Pencil or marker pen
- ■ Tape measure
- ■ Two 15mm isolation valves
- ■ 32mm waste pipework and fittings
- ■ Core drill (if you take the waste pipework through the wall)
- ■ Plumbers' mait
- ■ Slip-washer
- ■ Pump pliers
- ■ Adjustable spanner
- ■ Basin fixing kit
- ■ Screwdriver set

1 If you're replacing an old washbasin then the pipework should already be there. If there are no isolation valves then fit some now (remembering that they have an arrow on them indicating how they expect the water to flow through them), close them off and get the water back on again. Now jump to step 4.

2 If you're fitting a washbasin into a new position you need to run 32mm waste pipework into position, bearing in mind the maximum distance to the soil stack is 1.7m. The drop varies depending on the length of the pipe but normally horizontal runs should never drop by more than 18mm per metre.

Often the waste can go straight outside through the wall. Give yourself a bit of room for this by ensuring that the hole through the wall is a good 10cm lower than the outlet of the basin trap.

To get through an exterior wall hire a core drill and the appropriate core bit from a builder's merchant or hire centre. Not only do they save loads of time but they also create a much neater hole.

3 The hot and cold are in 15mm with the hot on the left and the cold on the right. To hide them behind the pedestal it's a good idea to set the pedestal and basin in place and mark the position of the pedestal with a pencil or marker pen.

Bringing the pipework up inside the pedestal is usually a lot more hassle than it's worth, but just behind the pedestal should hide them sufficiently.

You should aim for the waste pipework to come up between the hot and cold, so leave a gap sufficient for this.

Bring the hot and cold up about 40cm from the floor and fit isolation valves. Close them, turn the water back on and check for leaks.

4 You can now 'dress' the washbasin. First fit the waste by applying a ring of plumbers' mait (or clear sealant) to the top and pushing it through the hole – if your waste came with two washers then use the thin one here instead of the sealant or mait.

5 If you're using a pop-up waste you should have a thick washer and a base that now screws on to the piece you've just fitted through the hole.

The difficult bit here is getting the little protrusion on the side, which the pop-up mechanism fits into, to point straight towards the back of the basin. If you look at the photo you can see that when fully tight the pop-up adapter points to two o'clock. So if we loosen it off and rotate the entire waste – top and

bottom sections – anticlockwise by two hours (using the same clock analogy), then when we tighten it up again it will be fully tight at 12 o'clock – which by a lucky chance is exactly what we need.

The waste often comes with no washers. In these circumstances:

■ Apply a ring of plumbers' mait around the hole.

■ Add a slip-washer (slippery washers made from a hard plastic). These don't create a watertight seal – rather they allow you to tighten the main nut without smearing plumbers' mait everywhere.

■ Tighten the nut and remove the excess plumbers' mait.

6 Put the rubber washers on the base of the taps and put the taps through their holes. Hand-tighten them on the other side using the back-nut provided – see 'Replacing washbasin and bath taps' earlier in this chapter. Check the position of the taps. Usually they both point slightly inwards.

7 Some washbasins come with a separate plug hanging from a chain. If you have one of these you need to fit the chain at this stage. Just put the threaded bar through the hole in the basin and fit and tighten the nut at the back.

8 It's easier to fit a basin if you use flexi tap connectors, easier still if these are push-fit. Make your choice and fit them to the tap tails. Once hand-tight give them an extra three-quarter turn with a spanner.

TIP *If you fitted a monobloc tap you probably got flexi tap connectors with it. Often these are designed to fit the 12mm pipework used on the Continent, in which case you'll need to get a 12mm–15mm adapter.*

9 Fit the basin trap. The options available are:

Basin bottle trap – these are neat-looking but block up more readily. They can sometimes be too wide to fit inside the pedestal, but you can buy mini-bottle traps to get around this. You can also buy

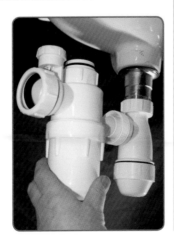

Left hand – Standard bottle trap with 'anti-vac'. Right hand – mini bottle trap.

bottle traps with an adjustable height, which can be very handy if your waste goes directly out through the wall.

TIP *If you're wondering what the bobble is on the top of the left-hand bottle trap, it's an 'anti-vac' attachment, which can be a very handy bobble to have. If when you empty the bath the waste starts gurgling and, in the worst cases, emits a noxious smell, what's happening is that the water racing through the pipework is trying to suck in air behind it. Sadly there is no air, so it sucks the water out of the basin trap instead, letting the smell of the sewers into your bathroom. The 'anti-vac' device prevents this by letting air into the pipework but keeping smells out.*

Basin P-trap –
more adjustments
can be made
with these. They
often come with
adjustable height,
ideal if you're
fitting a new basin
into an old position
and are having
trouble aligning the
waste pipework.

Basin S-trap
– good if the
pipework comes up
from the floor. Can
be adjusted to suit
the pipework.

**Straight-through
basin trap** – good
if your pedestal is designed so that the waste pipe runs down inside it. Can be awkward to get at once everything is fitted, and isn't as flexible as the S-trap.

Basin self-sealing trap – like the S-trap in terms of pros and cons, but they don't lose their water seal; ideal if this is a basin that's left unused for long periods.

You usually just need to hand-tighten the trap, but if in doubt give it a final three-quarter turn with pump pliers. Whatever you do don't start off with the pump pliers, as the plastic thread of the trap is very easily destroyed by the brass of the threaded waste; if it doesn't easily hand-tighten you're probably threading the trap.

10 With the washbasin completely 'dressed' it's time to fit it into position. First off, ceramic against ceramic is a very unforgiving scenario, so put a small amount of plumbers' mait around the top of the pedestal. If you haven't got any mait use plasticine; you just want something that'll act as a cushion.

11 Put the pedestal into position and put the basin on top. It's best to have a friend handy to keep everything steady.

12 Check that the basin is level. If it isn't, either use slivers of ply under the base of the pedestal or more mait on top of the pedestal

to get it level. Also ensure the basin is flat against the wall.

13 Mark the holes at the back of the basin. Sometimes these will be big ones, in which case you need to buy a 'basin fixing kit', but sometimes they're tiny little affairs, where rawlplugs and screws will suffice; occasionally both are provided, in which case aim to just use the big holes.

14 Measure the height needed for the hot- and cold-water pipework to reach the tap connectors, bearing in mind the pipework has to go into the fittings. The easiest way to do

this is to put a length of pipe into the isolation valve that's a little longer than needed, then put the end of the flexi connector next to it and mark the height you need to cut it back to.

15 Use the same method to determine the height for the waste.

16 Remove the basin and pedestal and drill your holes. If your basin has large holes then drill the appropriate-sized hole – most kits ask for a 14mm hole but you can get 10mm kits, which I prefer. If your basin has the smaller holes jump to step 18.

17 Use an adjustable spanner to screw the fixing kit bolts into the wall. They're fully in when the angled section that takes a spanner is hard against the wall.

 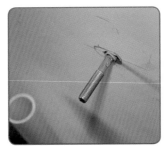

18 Fit the lengths of hot and cold pipe using your earlier measurements.

19 Fit the waste pipework. Note that if the old pipework was push-fit you'll need the same pipework and fittings now (take an old fitting down to a plumbers' merchant to discover what type of waste you have). If you can't find the same type go for solvent weld and use a 32mm compression coupling to join the two types of pipework together.

20 Put the pedestal back and fit the basin on top. If you used a basin fixing kit now's the time to put the nuts on the bolts coming out of the wall and tighten the basin into position, ensuring it's still level.

If your basin has the smaller holes fit an appropriate screw, making sure you use something to cushion the steel screw from the ceramic basin – either a small plastic or rubber washer or a plastic screw cover. Whatever you do don't over-tighten or you'll just break the ceramic.

With these basins with little holes it's also a good idea to add two blobs of sealant to the wall behind the basin to ensure it's secure.

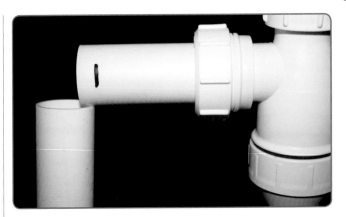

21 If you used a bottle trap or a P-trap you'll need to measure a short length of waste pipe to connect to the waste pipe coming up from the floor. Use the method described earlier, *ie* cut a length a little longer than needed, put it into the trap and then mark it against the pipework. Cut to the length you measured and connect the waste pipework up, hand-tightening the trap – this is usually sufficient, but give it an additional quarter-turn with a set of pump pliers if you're feeling pessimistic.

22 Connect the tap connectors to the hot and cold pipework and you're finished.

23 Open the isolation valves and check for leaks. If all is OK open the taps and let water run out the plughole. If all is still fine put the plug in and fill the basin. Let the water out and check again for leaks.

24 Pedestals often come with holes in the base so that they can be screwed to the floor. If you opt for this approach make sure you use rubber or plastic washers to separate the metal screw and the fragile ceramic. A more forgiving and possibly better approach is to use clear sealant around the base.

25 Apply a bead of sealant around the back of the washbasin.

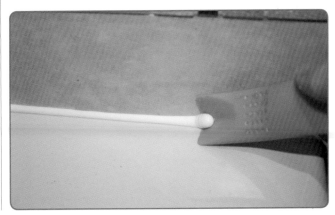

Fitting a new shower

Most people these days seem to prefer a shower to a bath, and most like a good strong shower. This is all well and good, and we'll be discussing the options on how best to achieve this. However, you need to bear in mind just how much water a strong shower can use. It's not unusual for a pumped mixer shower to use 14 litres per minute, which means that in five minutes they use enough water to fill a bath!

Courtesy of Bristan

Remember what I said at the beginning of this manual – that the UK gets less water per head of population than the Sudan? And things seem to be getting drier all the time. Unsurprisingly, therefore, the latest building regulations state that new homes must be designed to use only about 120 litres of water per person per day. If you're going to use a pumped mixer shower, flush your toilet once in a while and stick to this allowance then you'll barely have enough time to grab your loofah before you need to be out of the shower.

HIDING THE PIPEWORK

We'll have a look at the different types of shower you can buy in a minute, but whatever type you opt for they invariably look better if the pipework is concealed. However, hiding pipework in walls and behind tiles is fraught with difficulties if things start to leak. So let's just have a look at some of the issues:

- Wherever possible bring your pipework straight up or down the wall. This way you'll at least have some idea where the pipework is in the years ahead.
- Don't run electrical cable in the same channel as a water pipe. Electricity and water don't get on too well; electricity, water and a highly conductive metal pipe can generate real aggro.
- Use as few joints in the pipework as possible. Most showers need the pipes to come out of the wall at right angles so you'll need at least one elbow, but keep it to this – don't try bending plastic pipe as the bend you need is too tight.

- Avoid using any compression joints in the walls as they're more prone to leaking than push-fit and soldered joints. Sadly some showers need compression fittings in the walls but that's no excuse to go around adding to the problem.
- Ensure you test your pipework for leaks before you cover it up. Ideally use a pressure tester to raise the pressure to 1½ times the expected working pressure – 5bar is usually about right – and keep it at this pressure for at least an hour. If the pressure drops,

find the leak. You can usually hire pressure testers, if not locally then at least online. Note that if you're using plastic pipe you'll get a slight pressure drop as the plastic expands. So leave it for about 20 minutes, top up the pressure and then test for an hour.
- If you haven't got a pressure tester make sure you turn the water on in the pipework and leave it for at least a few hours to see if any leaks develop before covering it up.

Once you've tested for leaks cover the pipework. If you're using copper pipe use duct tape or use pre-coated pipe. Tile adhesive and concrete can corrode copper tube so it needs some protection. If you use plastic pipework fit a metal cover over it; this affords some protection and means you can find the pipework later using a metal/stud detector.

TIP *For some reason it's always tempting to run pipework down the middle of the shower area. However, when positioning a shower it's best to have the shower rail roughly*

in the centre of the shower area, with the shower unit itself off to one side. This sounds obvious and it is, but it's amazing how often it only becomes obvious after you've fitted the shower.

USING CHROME PIPEWORK FOR YOUR SHOWER

If I've scared you off the idea of hiding your pipework then you might want to run the shower water in chrome pipe mounted on the wall. If so, read 'Using chrome pipe' in Chapter 6 before you start.

FITTING AN ELECTRIC SHOWER

These are the easiest showers to fit from a plumbing point of view, as they only need a single cold-water pipe. The advantages of an electric shower are:

- Simple plumbing
- They heat the water instantaneously so you'll never run out of hot water for your shower, regardless of the number of teenage girls in the house.
- They're fairly economical.
- They supply water at mains pressure

The disadvantages are:

- They require a dedicated electrical cable from the shower directly to the main consumer unit. Modern showers can draw more than 10Kw of electricity, whereas your ring main can only handle 3Kw. As a result you must have the cable fitted by a Part P qualified electrician or, if you're replacing a shower, get the existing cable tested for the shower you're planning to fit. This is terribly important as death and fire can really ruin your day.
- In winter the cold water entering the house can be barely above freezing. To cope with this the shower has to slow down the flow of water through the heating element in order to get it up to temperature. This means you get a far poorer shower in winter than in summer. To get around this they now produce very powerful showers (10Kw+) but these also use more electricity.
- If someone opens a cold-water tap when you're mid-shower you'll notice a drop in pressure and, possibly, a drastic rise in temperature. To mitigate the latter point buy a thermostatic electric shower, ideally one that conforms to the TMV3 standard – especially if you have older folks or young children in your home.
- From a design perspective most electric showers are just plastic boxes stuck on your bathroom wall. They've recently come out with a few better designs but none are going to be hung in the Louvre any time soon.

Fitting tips

First off, read Chapter 6. This provides guidance on running pipework into place, which is 95% of the plumbing involved in fitting an electric shower.

a) The water should be supplied to the shower in 15mm pipework.

b) You need to be able to isolate the shower – and only the shower – so always fit an isolation valve to the pipework where it can be readily accessed. To make sure you don't restrict the flow use a full-bore lever valve.

c) Make sure you connect to the cold-water mains and not a stored water source. If you have stored water in your loft it's often easiest to connect your pipework to the mains cold water feeding the loft tanks and run it along the loft and down into the bathroom.

d) If you do run pipework through the loft make sure you insulate it.

e) If you have creatures in your loft consider using copper pipework and fittings, as rodents love plastic. If you want to go with plastic, raise it up and run it along the beams supporting the roof rather than across the loft floor.

f) Always flush new pipework before fitting the shower by opening the isolation valve and letting water run into the bath/shower for a few minutes.

g) Some shower manufacturers need the cold-water feed to be on the left side of the shower, some on the right. Bear this in mind if you're replacing a shower, as it will dictate if the new unit can fit directly over the space occupied by the old one. A number of companies now offer retro-fit products with swivel inlets that provide much more flexibility.

h) The point at which the shower connects to the pipework is often plastic these days. Although many of these can swivel to accept pipework that isn't angled perfectly they still tend to break after a while if the pipework has been forced into them, ie if you use plastic pipework with a bend rather than an elbow the pipe is constantly trying to straighten out, and this constant pressure will finally prevail. Also, don't over-tighten the connector or you'll break it.

i) Most electric showers require a 'commissioning' process, so make sure you actually read the manual.

j) When fitting the shower rail make sure you don't drill through the cold-water pipe or the electrical cable!

k) Ensure all the electrical work is completed and tested by a Part P qualified electrician.

FITTING A GRAVITY MIXER SHOWER

These are a little more awkward to fit as they need both hot and cold water, and these have to come from the same stored water supply (ie the same tank) or from two or more tanks joined together.

If you have a combi boiler or a high-pressure unvented hot-water system (see Chapter 2, 'How to identify your

system'), read 'Fitting a high-pressure mixer shower' below.

If you have a normal gravity-fed vented hot-water cylinder then you have a water tank in your loft to feed it, and the gravity mixer shower is a viable option for you.

There are two main types of mixer shower you can buy: exposed or concealed, the former sitting on the bathroom wall whilst the concealed is largely built into the wall with just the controls visible. The advantages of a gravity mixer shower are:

- They look nice! They're often a wonder in chrome and the concealed unit can be very minimalist.
- They're not affected by people turning other taps on and off.
- If the gravity aspect doesn't provide a good enough shower you can always fit a pump (see below).
- They're largely unaffected by the temperature of the mains cold water.

The disadvantages of a gravity mixer shower are:

- Because they use stored water you can run out of hot water, or even run out of water altogether.
- They're more difficult to install.
- They can be far more expensive to buy.

Courtesy of Bristan

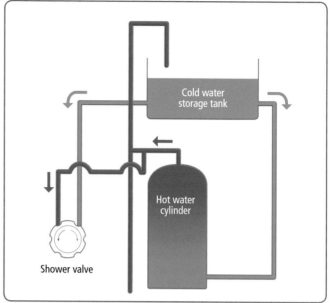

Cold water storage tank

Hot water cylinder

Shower valve

Fitting tips

Read Chapter 6 before attempting to install a mixer shower.

a) The diagram above shows how the water should be connected for a gravity mixer shower. Note that the hot and cold water start from different sides of the same tank and that the hot-water connection is the first branch off from the cylinder. Also note that the hot is going into the left side of the shower valve and the cold into the right.

b) Aim to run the water in 22mm pipework, changing to 15mm just as you drop down the walls to the shower. Read the earlier section on hiding your pipework.

c) Under no circumstances should you use cold-water mains pipework. It must come from the same tank that delivers the water to your hot-water cylinder. The reason for this is simple: the gravity hot water is under about 0.5bar of pressure, or less; the cold mains is usually at about 3bar. If you connect these different pressures to the same valve the shower will either run freezing cold or boiling hot and if you sneeze within 12ft of the shower it will change from one to the other.

d) The storage tank in the loft is usually only sized to deliver hot water. If you're going to take cold water from it as well then you either need a bigger tank or, far easier, you need to fit another tank alongside the first one. See 'Fitting a new cold-water storage tank' (Chapter 8).

e) Take the hot-water feed for the shower from as close to the hot-water cylinder as possible. You can buy special adapters that screw into the top of your cylinder and provide an outlet specifically for the shower. Alternatively you can cut into the pipe as it emerges horizontally from the cylinder and run the pipe from there. If you opt

for the latter always take the pipework down and then turn it up to go towards your shower. This is called a gravity loop and stops air being drawn into the pipe.

f) The hot water should always emerge from the wall on the left, the cold on the right. If this isn't possible many showers can be converted to accept the opposite feeds, but you'll need to read the manual for how this is done.

g) Fit full-bore lever valves to both the hot and cold water supply so that just the water to the shower can be turned off. Don't use ordinary isolation valves as they restrict the flow too much.

h) Many concealed mixer showers use a separate pipe to connect the shower valve to the hose. This pipe is also concealed in the wall and must be tested for leaks and given a protective covering before it's hidden under anything.

i) Always flush all new pipework with water before connecting the shower so as to remove any debris.

j) If you're fitting a concealed shower valve check your walls are thick enough to accept the valve and are made of suitable material.

Courtesy of Bristan

k) A number of bar mixer showers are only suitable for stud walls and not solid brick walls. Check that the shower is suitable for your type of wall before you buy, and see if there's a fixing kit available for that wall type.

TIP *When fitting lever valves, don't bother looking for any arrows on them to indicate the direction of water flow, as there aren't any. However, always try to fit them so that if the handle falls it closes the valve, ie is fails 'safe'.*

FITTING A HIGH-PRESSURE MIXER SHOWER

If you have a combi boiler or an unvented cylinder you have high-pressure hot and cold water, which by happy coincidence are the precise ingredients needed for a really good shower. Generally you can just buy a standard mixer shower as per the gravity mixer shower discussed above. However, there are slightly different fitting instructions.

Advantages of a high-pressure mixer shower
- The hot and cold water is at mains pressure.
- Some nice-looking chrome wonders and minimalistic designs are available.
- Works well in flats and bungalows.
- You'll never run out of water.

Disadvantages of a high-pressure mixer shower
- If you have an unvented cylinder there really aren't any, aside from the tiny risk of running out of hot water. However, if you have a combi boiler there are a few:
- A combi boiler heats up the water instantaneously, but it can only do this effectively up to a certain flow rate, which will decrease in winter when the initial temperature of the mains cold water drops.
- If someone opens another hot- or cold-water tap whilst you're having a shower there's a very good chance you'll know about it. As such always opt for a 'thermostatic' shower.
- You might live in an area where the mains pressure is fairly low so always get this checked beforehand, as it might restrict what you can and can't do – body jets, for example, might be out of the question. This also applies to unvented cylinders.

Fitting tips
In essence fitting is the same as for a gravity mixer shower, except that you aren't taking the water from a tank. In fact you can usually connect the new pipework for the shower directly to the hot and cold pipework feeding the bath.

A flow restrictor is usually supplied, and the fitting instructions will state that this should be installed if you have a combi boiler. Although there's a temptation to ignore this guidance, as you don't want to restrict the flow to your shower, you should fight it – the restrictor's there to ensure that you still get a good shower in the winter. Omit it and your midwinter showers will be cool and meagre.

FITTING A SHOWER PUMP

Whilst gravity mixer showers look good and often work well, they don't always supply enough water for a really good shower, especially if the cold-water storage tank is only just higher than the shower head. To resolve this you can add a shower pump.

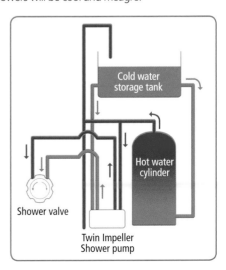

Cold water storage tank

Hot water cylinder

Shower valve

Twin Impeller Shower pump

Generally speaking you want to purchase a 'twin impeller' pump providing about 1.5bar of pressure. If your shower has body jets the manufacturer will probably recommend a 3bar shower pump, at which point the difference between showering and riot control becomes open for debate.

In houses, and particularly flats, where it's not possible to site the water tank in the loft and the stored water is consequently at or below the height of the shower head, an ordinary pump won't work and you'll need to buy what's called a 'negative head' pump. This can be a particular pump model, but often it's an 'add-on' kit for a standard pump.

> ⚠️ **WARNING**
>
> You should never, ever, pump mains-fed hot or cold water. Only ever fit a pump to water that comes from a storage tank.

1 The ideal place for a shower pump is at the base of your hot-water cylinder. Some get fitted in lofts, but this leaves them exposed to freezing in winter and sometimes invalidates the manufacturer's warranty. If you do fit one in the loft remove the insulation from underneath it and try to ensure it's snug and warm whilst still leaving enough air movement so that the electric motor doesn't overheat – a difficult balancing act at the best of times.

Most pumps have little rubber feet, which sometimes have to be stuck on first. These reduce the noise when the pump's running, so don't forget them. If the pump is still noisy you can try fitting an old mouse mat underneath the legs to reduce the vibration still further.

2 Turn off the water to your shower at the isolation valves and open up the shower valve to drain it. Put the hose as low as possible to get as much water out as you can.

3 The pipe taking hot water into your shower should run directly from, or very close to, the hot-water cylinder. Identify this pipe, which should already have a full-bore isolation valve

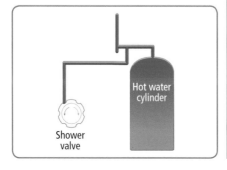

attached. Redirect the pipe from the isolation valve to the inlet valve of your shower pump and from the shower to the outlet port of the pump.

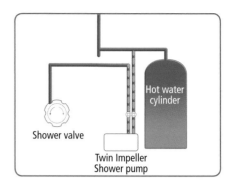

Make sure both inlet and outlet pipework have full-bore lever valves on them so that you can easily remove the pump if necessary – some manufacturers provide flexible hoses that already

contain isolation valves. The inlet and outlet ports on the pump are usually clearly marked and it doesn't matter which side is used for hot or cold.

The inlet connector usually has a little filter at its base. Make sure this is in place before you tighten it up.

Always flush the pipework with water to remove any debris before fitting into the pump.

4 Cut into the cold-water pipe serving the shower after the isolation valve and redirect the two pipe ends to the pump; one will be the inlet, the other the outlet – it's a good idea to mark which is which.

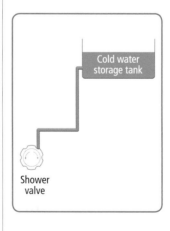

Fit lever valves to both pipes so that the pump can easily be removed if required, and connect the pipes to the appropriate ports on the pump, remembering to flush the pipework first and check that the inlet filter is in place.

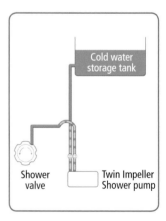

TIP *If your pump starts acting up the usual culprit is debris clogging the inlet filters. To fix this, isolate the pump from the electricity supply, close the four isolation valves, undo the inlet valves and remove the filters. Give them a good clean and fit everything back together again.*

5 Your pump should now be connected to four pipes and be resting securely on a solid surface. Open the isolation valves and let water run through the pump and out of the shower hose. This ensures the pump is filled with water.

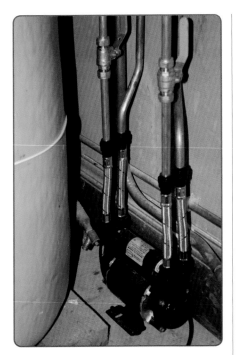

6 The pump now needs to be connected to an electrical supply via a fused spur. It's best to get a qualified electrician in to do this work for you.

FITTING A POWER SHOWER

A power shower is really just a mixer shower with a pump built into it. They seem to have gone out of favour recently, presumably due to their rather bulky appearance. The advantages of a power shower are:

- The water is pumped, so you get a really powerful shower.
- Because the pump is built in there's only one thing to buy, so they can work out cheaper than a pumped mixer shower.
- The pump power is often adjustable, so you can save water when needed and blast yourself across the bathroom when the feeling takes you.
- They're not affected by anyone else in the house using another water source at the same time.

The disadvantages of a power shower are:

- They use a lot of water if you're not careful
- They take their water from a stored source, so you can run out of water.
- They require a positive head, *ie* the water storage tank needs to be sited above the height of the shower head, which isn't always possible.
- You need to involve an electrician, as they need wiring in.
- They're often unappealingly large plastic boxes stuck to your bathroom wall.

Fitting tips

a) If you have a combi boiler or an unvented hot-water cylinder you can't fit a power shower – they must take their water from a stored, gravity-fed supply.

b) The pipework is the same as for a gravity-fed mixer shower (see above).

c) Get a Part P qualified electrician in to make the electrical connections.

DIGITAL SHOWERS

These are effectively a pumped or gravity-fed mixer shower. The only difference is that the internal controls have been separated out and can now be tucked out of sight – usually in the loft. The result is that they look wonderfully minimalist and installation is less disruptive. Many also allow you to adjust the water flow so you can run them in 'eco' mode one day and immerse yourself in luxurious hot water the next. On the downside, they can be very expensive and you will have to get an electrician involved.

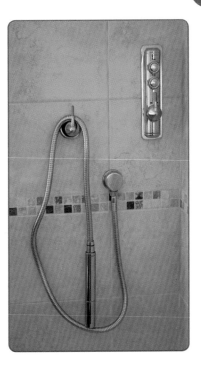

PUMPED ELECTRIC SHOWERS

I know of only one shower that fits in this category, the Mira elite. It's like a cross between a power shower and an electric shower in that it takes in just cold water and heats it up as it passes through the shower element, as per a standard electric shower. However, it takes its cold water comes from a stored supply, and can therefore be pumped, as per a power shower.

I've fitted a few of these over the years and they've all been in tiny little out-of-the-way villages where the cold-water mains pressure is poor. If this is the situation in which you find yourself this might be a good option for you.

To fit, you just take a cold-water supply from the cold-water storage tank in your loft (as per the mixer shower mentioned earlier) and then fit an electrical cable as per a standard electric shower.

Fitting a bidet

For some reason, people in the 1970s awoke to the realisation that they didn't possess a sparkling rear end and that toilet paper was woefully inadequate. Overnight everyone suddenly needed a bidet in their bathroom and it took another 20 years before people realised

that, actually, they probably didn't. That said, there are sound reasons for a bidet and some people out there might want to fit one.

The general fitting instructions are pretty much a mix between fitting a toilet and fitting a washbasin, both of which we've covered, so we can probably dispense with a step by step guide. However, there are specific regulations that apply to bidets so let's at least have a look at these:

- Because of the nature of a bidet, the water companies are very anxious that the contents of a bidet don't make their way back into the water supply. If your bidet has a hose attachment or the water shoots up from the base then the cold-water supply must come from a separate storage tank that only supplies the bidet. The hot water needs to come from a separate hot-water heater, supplied with cold water from this 'bidet tank'.
- The base of this tank needs to be at least 15mm higher than the maximum height of the bidet hose.
- If the bidet taps discharge their water down into the bidet life is a bit easier. Because the water will overflow the bidet bowl before it reaches the tap outlets it's not considered a real danger. As such you can use the same hot and cold water as everything else in the bathroom, providing the bidet is lower than the bath.

Although it's not specifically required, I'd also recommend fitting double-check valves to the hot- and cold-water supply just before the bidet taps.

Fitting a shower tray and cubicle

These vary from manufacturer to manufacturer so we're not going to step through the installation process – you just follow the manufacturer's installation guide. However, there are a number of important points you'll need to bear in mind:

- A quadrant shower tray and cubicle takes up far less space than the standard square or rectangular tray. They also do offset quadrants.
- The tray will leak if it moves so make sure the floorboards underneath are solid, and if you have any doubts take them out and fit a sheet of 19mm plywood instead.
- You can buy feet for the shower tray, usually called a 'riser kit'. It's often a good idea to supplement these with blocks of wood to ensure the tray doesn't move.
- The waste will normally dictate how low the shower can go, and the floor joists will normally dictate how easy it's going to be to run the waste under the floor.

- If you're going through joists drill holes rather than notching them, as notches weaken the joists far more.
- If you must go through a joist the maximum hole diameter is one-third the depth of the joist or, if notching, quarter the depth. That said, try not to go through joists at all, as doing so will always weaken the floor to an extent.
- The shower needs a 40mm waste pipe and should be within 3m of the soil stack.
- The tray needs to be perfectly flat, so take time to ensure this.
- It's best if the tray is recessed into the wall a little and then stuck back using clear sealant. You usually only need to remove the plasterboard on a stud wall, or the rendering on a solid wall to achieve this.

- You can buy trays with 'upstands', which are little ridges around the edge. You install the tray so that these are flush with the wall. Then you tile down into the tray. As a result it's almost impossible for these trays to leak. Cheaper trays are usually flat-topped, but it might be worthwhile paying a little more and getting one with a number of 'upstands'.

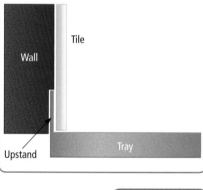

- Once the tray is in, tile down to it, leaving a gap of about 3mm. Fill this gap with sealant and not grout.
- Fit the cubicle to the tiles. Never fit the cubicle first and tile afterwards.
- Always use clear sealant to create a watertight seal right down the back of the wall profiles. Also run a ring of sealant around any screw holes, and add a good dollop where the base of the wall profile meets the tray and wall. Bizarrely this is rarely mentioned in the fitting instructions, which possibly explains why it's the most common cause of a shower tray leaking.

- Only ever seal the outside of a shower cubicle. This allows any water that gets into the frame of the cubicle to leak out into the tray rather than stagnating in the frame or running out on to the carpet.

Swapping your bath for a shower tray

Whilst most of us prefer a shower to a bath, we nearly all have to clamber into the bath to take our shower. For kids and for the elderly this can be a right pain, yet we seem extraordinarily reluctant the take the bath out and put a shower cubicle in its place.

The reality is that it only takes a couple of days to get a bath out and a shower tray in, and even less time to take the shower out and put the bath back in. So why make life difficult?

You can buy a shower tray with the same footprint as the bath or you could put in a smaller cubicle and make your bathroom look a whole lot bigger. The only issues to consider are:

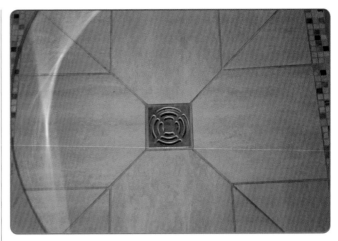

- The shower waste will usually be lower than the bath waste, so consider how you're going to achieve this before you start the project. The usual options are to fit a riser kit to the shower tray or redirect the pipework to join the main waste lower down – which isn't always possible.
- The hot and cold pipework for the bath needs to be capped off as low to the ground as possible – ideally under the floor.
- Check that there are no other water pipes under the bath as these will all have to be moved.
- The underside of the bath is usually not plastered or rendered, so you'll have to apply some bonding to it before you can tile.

Wet rooms

The wet room has come into its own recently and it has to be said that they can look beautiful when done right, and there's something weirdly liberating about showering without a bath or tray to limit you. On a more practical note they are, of course, ideal for the disabled in that they allow wheelchair access to the shower.

There are two approaches you can take: you can either tile the entire floor or you can use a waterproof vinyl. The tiling approach is a lot more complicated and prone to disaster if done wrong. Vinyl wet rooms used to end up looking like a cross between the public baths and a hospital operating theatre but these days they come in a much better range of designs and have lost the 'sanitized' look they used to have.

If you're contemplating a wet room you need to bear the following in mind:
- How are you going to get the shower waste out of the room? By definition it has to be lower than floor level so you're either going to have to run it under the floorboards or channel out a section of concrete floor that can accept a 40mm pipe. If you've looked long and hard and just can't see how you can run waste under the floor, you might want to consider using a pump to lift the waste to the main soil stack. These are a bit noisy but they will allow you to fit a wetroom when gravity is saying no.
- You need the floor to be perfectly flat. To achieve this on a concrete floor you're best off applying a 'self-levelling' compound. This is a runny latex cement that you pour over the floor. You use a trowel to get it into all the corners and work it until it's roughly flat, then just leave it to find its own level. To get a wooden floor level you might have to take up the floorboards and replace them with sheets of ply or tongued and grooved chipboard flooring.
- The easiest way to create the tray area is to buy a 'tray former'. These are either plastic or wood and create the shape of the tray. You recess them into the floor and then apply your flooring over the top.

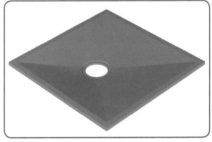

- If you're going to tile the floor you'll need to make sure it's completely waterproof. To do this you need to 'tank' it first, which involves applying a waterproof latex paint or covering over the entire floor and at least partway up the walls. This isn't cheap but is essential.
- If you have a concrete floor tiling is a viable option, but if you have a wooden floor I'd personally veer away from tiles, unless you're prepared to spend the time and money strengthening them first. If your floor creaks like a crazy thing then think about removing the floorboards and replacing with waterproof caberboards or 19mm marine ply. Once the floor is stable, put something like hardibacker – thin concrete sheets – over the top to get everything absolutely solid before tanking the room. Unless you're a very good tiler I'd also leave the tiling to a professional.

Most vinyl flooring can be bought with installation included; personally I'd take them up on this.

If a wet room sounds far too complicated go for the compromise, which is a low-level shower tray. These are far easier and far cheaper to fit.

12 CENTRAL HEATING AND HOT WATER

Bleeding a radiator 173

Maintaining your central-heating system 173

Keeping your central-heating system clean 177

Removing a radiator 179

Fitting a new radiator 181

Fitting new radiator valves 184

Fitting a heated towel rail 185

Common central-heating problems 186

Replacing the central-heating pump 189

Replacing the motorised valve 191

Fitting a new hot-water cylinder 192

Replacing an immersion heater 197

Boilers 199

In Chapter 2 we talked about how to identify the plumbing systems in your home. Having read those chapters you should be pretty confident about what type of CH system you have, but if there are any lingering doubts call a Gas Safe registered engineer and ask them to have a look. It oughtn't to take more than half an hour and I'd be surprised if they charged more than £40. It's a one-off requirement, and if nothing else should confirm what you'd already figured out for yourself.

The other task covered in Chapter 2 was how to drain down and refill your CH system. Knowing how to do this is a prerequisite for many of the topics we'll discuss here so you might want to reread that chapter.

Let's break into the subject of central heating nice and gently.

Bleeding a radiator

As you heat water up it releases the air trapped within it. In the meantime the general corrosion within the system is releasing a variety of gases. In an ideal world these gases would work themselves out of your central-heating system via vents and automatic air-release valves. Alas, we don't live in an ideal world so they accumulate in the radiators.

The easiest way to tell if this is happening is to feel the radiator. If it is hot at the bottom but cold at the top it's almost certainly filled with air and needs 'bleeding'. The first task is to find the bleed valve:

■ These are always towards the very top of the radiator.
■ They can be on either side.
■ They're usually fairly obvious but can be located behind the back of the radiator and hidden from view behind a plastic cover.
■ Most heated towel rails have the bleed valve right at the top.
■ A double-panel radiator will often have two bleed valves, one for each panel.

Having found the valve you need to make sure conditions are ready for 'bleeding' – *ie* the CH system is switched off, so the pump isn't running, and all the radiators are cold, or at least cool. If the pump is on you can get all sorts of strange effects when you try to bleed the radiators. If the water is hot when you bleed the radiator it will contract as it cools and draw air into the system – which is what you've just been trying to get rid of.

You're now ready to 'bleed'.

Tools and materials
■ Bleed key
■ Tissue paper

1 The water that comes out of a radiator can be very dirty, so hold a piece of tissue under the valve.

2 Turn the bleed key anticlockwise to open the valve. Only turn the valve a few turns and don't remove it altogether.

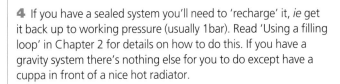

3 You'll hear air escaping from the valve. As the water reaches the top of the radiator this will turn into a series of wet splutters. This is the time to close the valve again, whilst catching the excess water in your tissue paper.

4 If you have a sealed system you'll need to 'recharge' it, *ie* get it back up to working pressure (usually 1bar). Read 'Using a filling loop' in Chapter 2 for details on how to do this. If you have a gravity system there's nothing else for you to do except have a cuppa in front of a nice hot radiator.

Maintaining your central-heating system

The vast majority of us give our central-heating system little or no regard; right up to the moment it starts to act up. However, with just a little time and thought it's possible to keep it all in good working order.

When first installed everything about your central heating was clean and shiny; even the water it contained was clear, possibly not crystal clear but at least it was a liquid. Over the years things happened. The metals started to corrode each other and the oxygen in the water encouraged this. Eventually the water turned into thick, black gunk – the central-heating equivalent of arteriolar sclerosis – and this gunk started to line the inside of the radiators,

the boiler and the pipework. Not only does this reduce the life expectancy of your CH system but it slowly makes it less efficient.

This is fine if you have money to burn, because that's exactly what you're doing. For the rest of us it's not such a good idea, but what can we do about it?

First off, how do you tell if your CH system is full of gunk? Well, there are a few telltale signs:

■ When you bleed your radiators is the water very dirty?
■ Do you have radiators that once got hot but now don't?
■ Do your radiators have cold spots in them?

If the answer to any of these is 'yes' then your system could probably do with a good clean, which is known in the trade as a 'powerflush'.

POWERFLUSHING THE SYSTEM

A powerflush is where you add chemicals to your CH system to start breaking down the dirt and gunk. You then use a large pump to dislodge any remaining dirt, before dumping the dirty water to waste and replacing it with lovely clean water. Things to bear in mind before you start are:

■ The chemicals can be bought from your local plumbing merchants or DIY store, whilst the pumping equipment can be hired from most building/

> ⚠ **WARNING**
>
> Do you have a Primatic hot-water cylinder? The only real way of identifying these – aside from looking for a label with 'Primatic' written on it – is to check the number of tanks in your loft. If you have only one large tank but know you don't have a sealed or unvented CH system, then you might have a Primatic cylinder.
>
> Primatics work by separating the CH water from the domestic hot water by an air bubble within the tank. God alone knows how these were ever regarded as a good idea, but they were. Because there's only this air bubble separating the two systems you shouldn't powerflush your CH, nor should you add any chemicals to the CH water. Personally, I'd jump to 'Fitting a new hot-water cylinder' and replace a Prismatic with a standard indirect hot-water cylinder as soon as your finances allowed.

plumbing merchants and general hire stores.
■ The main requirement of a powerflush is time; there's no point trying to do it all in 15 minutes – you need to set aside an entire day to do it properly.
■ What kind of CH system you have determines the steps needed to complete a powerflush. If you have a sealed, pressurised CH system (a 'system' boiler or a combi) then you're better off fitting the powerflush to a radiator. If you have a 'sealed system' jump to step 6.

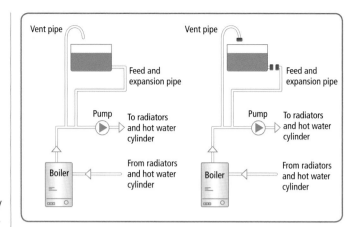

■ If you have an open-vented system, *ie* you have a small tank in your loft (the feed and expansion tank) feeding water into your CH system, then you have two openings in your CH system that need to be blocked off first – the vent pipe and the feed and expansion pipe. If you don't do this you're going to get water all over the place.

Tools and materials
■ 22mm and 15mm push-fit stop-ends
■ Flat-headed screwdriver
■ Garden hose
■ Jubilee clip for hose
■ 22mm and 15mm pipe cutter
■ Tub of sludge remover
■ Powerflusher
■ Rubber mallet
■ 15mm push-fit straight coupler

1 Turn off the cold water to the feed and expansion tank, either by fitting an isolation valve just before the ballcock or by putting a length of wood across the top of the tank and tying the ball valve arm to this wood so that it won't open. The isolation valve is always the better option.

2 Drain the tank by opening a drain-cock on the CH pipework downstairs and letting the water run out of a hose and into an outside gully – don't drain the whole system, just let the tank drain

TIP *Most of the gunk you're trying to clean out is magnetic. With this in mind you can now buy magnetic powerflushers. I haven't used one yet but it makes sense to me to remove magnetic debris with a magnet rather than with lots of water, so you might want to check out this approach.*

completely then count to 60 and close the drain-cock. Now that you have a dry tank you might want to take the opportunity to clean it out.

3 Cut into the feed and expansion outlet pipe using a pipe cutter.

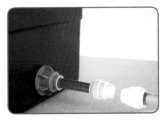

4 Fit a stop-end to each open pipe end.

5 Cap off the CH vent pipe with a 22mm push-fit stop-end. This is the pipe that terminates over the smaller tank of the two up in your loft.

6 Buy your chemicals. There are many manufacturers selling these and they go by a number of different names but most have the words 'sludge remover' on the label somewhere. You often see the words 'system cleaner' as well, but those chemicals are usually designed to clean out newly installed CH systems, not old dirty ones.

Read the labels carefully when buying and avoid ones that can't be used with aluminium heat exchangers, for the simple reason that the odds are you haven't got a clue what the heat exchanger in your boiler is made of.

Note that some chemicals are designed to be added to your system for just a few hours whereas others can be left in for weeks. It's going to make life easier to opt for the quicker-acting chemicals, although both will work. One litre is usually enough for ten radiators.

7 Hire your powerflush machine. Try to get one with a reverse flow and check that it's in decent condition and has the full complement of hoses and adapters before you hire it. Don't hire one that doesn't have the user manual with it; most machines are similar but you need to follow the user guide.

8 Decide the best place to connect the powerflush to the system. On a sealed system opt for the smallest radiator that is near a source of water and a waste outlet, *ie* the kitchen or bathroom radiator – see 'Removing a radiator' later in this chapter. Once removed fit the powerflush hoses to the radiator valves. Using a radiator is the easiest place to fit the powerflush and is the location I'd recommend.

As an alternative you can use the CH pump. On either side of the pump are two isolation valves. Remove the pump (see 'Replacing the central-heating pump' later in this chapter) and fit the powerflush to these valves.

Some powerflush units come with an adapter that allows you to remove the 'face' of the pump and fit the hoses into the pump body.

If you're removing the pump make sure you remove the power to it. The easiest way is to turn off the electricity to the pump, remove the live, neutral and earth from the pump itself and insulate the wires. If you're not happy doing this then fit the powerflush unit to a radiator instead.

9 Follow the instructions that came with the powerflush to set it up. Note the following:

- An overflow is fitted but it's often better to sit the entire powerflush unit in a large bucket – they often come inside a bucket when you hire them.
- Bearing in mind that you might get leaks site the unit in a bathroom or kitchen, *ie* on a tiled floor, not your best shag-pile carpet.
- You need access to water and waste outlets. It's often best to take the water from the outside tap and run the waste to a gully outside. If that's not possible run the waste into the toilet pan, but make sure it won't fall on to the floor at inopportune moments.
- To ensure the CH pipework is all open you need to set any motorised valves to open. You can do this manually by pushing the little lever to one side and pushing it into the 'lock' position;

some valves have slightly different mechanisms but they all have a manual setting (MAN), which is always an 'open', so check them out or jot down the make and manufacturer and look them up online.

10 When you connect the powerflush to the CH make sure the valves on the powerflush are closed before you open the radiator or pump valves.

11 Open up the valves on the powerflush and let the container fill above the 'minimum' mark before turning the pump on.

12 To start off it's best to run the existing water in the CH system to waste whilst filling the powerflush with clean water. This gets rid of the bulk of the dirt. Once the water is looking fairly clean stop running to waste, turn off the water and add the chemical cleaner.

13 If possible turn the boiler on to get the water temperature up to about 50°C. Don't leave it on – you just want to warm the water, so turn it on for 15–20 minutes at most.

Open all the radiators and let the powerflush circulate the warm water for at least an hour. You need to open the valves on both sides of the radiator. One side will have a cap that you can turn to allow this.

For the other side (the lock-shield) you'll have to remove the cover and turn the head anticlockwise with a small spanner. As you do this, count the number of turns you make until the valve is fully open. Jot this down so you can set the lock-shields back to how you found them.

If you have a powerflush with a reversible setting, remember to reverse the flow every five minutes or so.

14 Turn off all the radiators except one by turning the valve with the head anticlockwise. You're now cleaning just this one open radiator. It's a good idea to give the radiator a few taps with a rubber mallet to loosen up any debris inside. (You can buy a handy little tool that fits on the end of a drill and does the job of the rubber mallet by agitating the radiator for you.) Reverse the flow a few times as this also helps dislodge any dirt.

Give this radiator at least ten minutes – the longer the better really. Once you're happy that it's clean open up the next radiator then close this one so that once again everything is flowing through a single radiator. Repeat this step for every radiator.

15 When you've finished the last radiator leave it open and start running the CH water to waste whilst replacing it with fresh water.

If the powerflush is going to overflow now's the time it will do it, so monitor the water level in the powerflush and if necessary adjust the amount of fresh water entering it.

16 When the water starts to run clear open the next radiator and close the first one. Now run this water to waste until this radiator also runs clear. Repeat for every radiator.

17 Now open all the radiators whilst still running the water to waste and refilling the powerflush with fresh water. This is the last stage, so wait until the water is looking beautifully clean and then stop.

18 Isolate the powerflush from the central heating by closing the radiator or pump valves and pack it away.

19 Refit the pump or radiator – see the appropriate sections further on in this chapter.

20 For an unvented system take the stop-end off the vent pipe, then remove the two stop-ends from the water tank outlet and replace with a push-fit straight coupler.

21 Set any motorised valves back to automatic by lifting the lever and letting it find its own place in life.

You now need to add inhibitor to the system. For details on how to do this read the next section, 'Keeping your central-heating system clean'.

22 Finally, refill the CH. For a sealed system this means opening the filling loop and letting the pressure rise to 1bar, topping it up as you bleed the radiators – see 'Bleeding a radiator' earlier in this chapter. For a vented system it means opening the ball valve, either by releasing the string holding it up or opening the isolation valve.

With water once more entering the system bleed all the radiators and reset the lock-shields to their previous settings following the notes you made in step 13.

Keeping your central-heating system clean

Now that you've gone to all the trouble of getting the water in your CH system clean, you need to keep it that way. There are two approaches you can take and ideally you'll do both:

ADDING INHIBITOR TO THE SYSTEM

Inhibitor is a chemical that slows down the rate at which corrosion takes place within your CH. It doesn't stop corrosion altogether and you need to top up the levels once a year.

Lots of different companies make inhibitors, and usually the tub or tube supplied is enough to treat ten radiators. Most come with a detachable label for you to jot down the date you added the inhibitor. Stick the completed label next to your boiler so everyone knows what was used and when.

SEALED SYSTEMS

For sealed systems you need to buy inhibitor concentrate. To use this:

Tools and materials
- Utility knife
- Sealant gun
- Tube of CH inhibitor concentrate

1 Cut the end off the tube. Attach the adapter that seems to fit into your bleed valves best and fit the tube into a sealant gun.

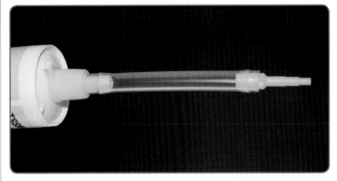

2 Turn off both valves to the radiator by turning the head clockwise; for the lock-shields you'll need to remove the cover and turn the revealed head clockwise with a smaller spanner. Open up the bleed valve using the key and let the water run out into a bowl.

TIP *If you've just powerflushed the system and you'd connected the flush unit to a radiator then this radiator should be empty, which makes it ideal for adding the inhibitor to.*

If this isn't possible it's often easier to add inhibitor to a towel rail, as the bleed valves are right at the top and can even be removed completely – but only after you've closed both radiator valves.

Otherwise pick the biggest radiator in the house, as you need some space to get all the inhibitor in. Even then some will drip down the radiator, so put a cloth underneath.

3 Once the water stops coming out remove the bleed screw completely. Keep an eye on this screw as it's ludicrously easy to lose.

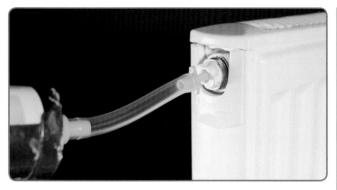

4 Fit the adapter into the bleed hole and slowly squeeze the liquid into the radiator. Once the radiator is full the inhibitor will just start to pour down the side. If this happens you might want to stop, jump to step 5, and then add the rest of the inhibitor to another radiator by repeating steps 2-5.

5 Replace the bleed screw. Open up the radiator valves and recharge the system with the filling loop.

If you think this is all a right palaver you might want to buy a magnetic cleaner (see below). Not only do these keep the CH system clean but they also make adding inhibitor much easier.

GRAVITY-FED SYSTEMS
For gravity-fed systems you need to pour a tub of the liquid into the feed and expansion (F&E) tank, but first this tank needs to be emptied. If you've just powerflushed the system (see earlier) the tank is already empty. In this case, just connect up the outlet pipe as directed and pour the inhibitor into the tank. Otherwise you need to empty the tank first. To do this:

Tools and materials
- Garden hose
- Jubilee clip to fit hose
- Wood and string (if your ballcock doesn't have an isolation valve)
- Tub of inhibitor
- Bleed key

1 Close off the water to the F&E tank either by turning it off at the mains, turning off the isolation valve just before the ballcock or by putting a piece of wood over the tank and using a piece of string to tie the ball valve to it.

2 Find a downstairs radiator that has a drain cock and is close to a door. Fit a hose to the drain cock and run the hose outside to a drain or gully.

3 Open the drain cock and let the F&E tank empty.

4 Once empty count to 120 then close the drain cock and remove the hose.

5 Pour the inhibitor into the now empty F&E tank.

6 Let water back into the tank. You shouldn't need to bleed the radiators, but you might want to check those upstairs just to be sure. Again, you you'll find adding inhibitor to the system much easier if you add a magnetic cleaner to your CH system.

FITTING A MAGNETIC CLEANER TO YOUR SYSTEM
As already mentioned, the vast majority of the gunk inhabiting your CH system consists of corroded metal. The great thing about this is that the metal is magnetic, and since it tends to bind to pretty much all the other debris in the system it means that nearly all the detritus that kills off your CH can be removed by a magnet.

A number of companies now make magnetic cleaners and some add other bits and bobs to the basic mechanism. They aren't difficult to fit: just drain your CH, cut into the pipe at a convenient location – usually on the return pipe below the boiler – fit the cleaner with its isolation valves, refill the system ... and that's it! However, there are many variations on this theme so make sure you follow the manufacturer's instructions. Once fitted you just need to remove and clean the magnet occasionally – every three months initially, but once a year thereafter should be sufficient.

Most of these cleaners can also be used to add inhibitor to the system. Just isolate them at their valves. Open them up. Drain the contents of the container. Add your inhibitor. Close them up again. Open the isolation valves and hey presto. They're often worth buying for this reason alone.

Removing a radiator

Removing a radiator is simple and straightforward. In fact the only awkward aspects are the weight of the radiator and the fact that the water inside it is usually filthy and just longing to pour itself all over the carpet.

TIP *All radiators come with two valves. One is called a lock-shield and the other either a 'wheel head' or a 'thermostatic radiator valve' (TRV). The wheel head and TRV are used to turn the radiator on and off. The lock-shields, on the other hand, are set for each radiator when the CH system is first installed.*

A typical lockshield valve

They control the resistance to water within the radiator and as such ensure that all the radiators in the house get hot, as opposed to one or two getting desperately hot whilst the rest remain stubbornly lukewarm or just plain cold. With this in mind, the lock-shields are usually covered by a lid that stops you accidentally turning them and altering their setting.

When you do need to close the lock-shield completely – for instance when changing a radiator – it's important that you count the number of turns you make so that you can set it back to its original setting afterwards.

Whilst you can remove a small radiator by yourself it's much easier if there are two of you, so rope in a friend.

Tools and materials

- Dustsheets
- Small adjustable spanner
- Larger adjustable spanner
- Set of Allen keys (depending on type of radiator valve you have)
- Tissue paper/small bowl
- Bleed key
- Pump pliers
- Water vacuum or tray that fits under the radiator
- Two short lengths of 15mm copper pipe
- Two 15mm nuts and olives for the radiator valves (¾in for older radiators)
- Two 15mm stop-ends

1 Roll back carpeting from under the radiator so that if you do get water on the floor it's not going to ruin anything. Put a dustsheet under the radiator. Decide what route you're going to use to get the radiator out of the house and cover this with dustsheets.

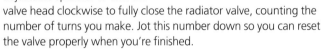
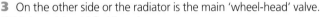

2 Take the cap off the lock-shield valve and use a small adjustable spanner to turn the valve head clockwise to fully close the radiator valve, counting the number of turns you make. Jot this number down so you can reset the valve properly when you're finished.

3 On the other side or the radiator is the main 'wheel-head' valve. If this is a plain old-fashioned valve then turn the head clockwise until it's closed. Now jump to step 6.

4 Today this valve is more often a thermostatic radiator valve (TRV). You can tell if this is the case because there'll be numbers on it indicating the temperature – 5 is high, 1 is low, 0 is closed completely.

When these valves were fitted they should have had a plastic cap that went with them. If you have this cap remove the valve 'head' by undoing the nut at the base of the head. Some unscrew by hand, others need an Allen key.

Once the head has been removed, screw the cap down hard on the exposed valve. This closes the valve altogether and you're ready to jump to step 6.

5 If you can't find a cap for the valve then you can close it down completely by turning it clockwise until first a star (or snowflake) symbol appears, and then a zero.

This has closed the valve, but there's a caveat; if the temperature in the room drops towards freezing the valve may open itself up again. It shouldn't do this if you have it set to zero, but I have known it to happen and it usually occurs at about three in the morning. To avoid flooding your home in the small hours use the cap mentioned in step 4, or try to find something to cap off the end of the valve once you have the radiator off.

In truth, if you're going to leave the radiator off for any length of time you should always cap off the valves. If you have inquisitive kids in the house don't even leave the room without doing this first. To cap off the valve take a short length of copper, put a nut and an olive on it and screw it into the radiator valve. Now fit a push-fit stop end onto the other end of the copper tube.

6 The radiator is now isolated from the rest of the CH system, and you can now release the water pressure inside it. To do this

take some tissue paper or a small bowl and hold it under the bleed valve, open the valve and let the water run out.

7 When the water stops running fully close the bleed valve. If water doesn't stop running out it means that one or both of the radiator valves aren't fully closed. Check them. If you're convinced they *are* fully closed then it looks like one or both has failed and needs replacing – see 'Fitting new radiator valves' later on in this chapter.

8 This is where it can potentially get very messy, so place some old towels and a bowl under the radiator. Have a replacement bowl nearby and, if you have one, set a water vacuum ready to go.

9 Hold the valve steady with a set of pump pliers and open the radiator tail by turning the nut anticlockwise (anticlockwise when looking at it from the radiator). You might want to protect the valve by putting a wad of cloth between it and the pliers. It's very important that you support the valve. If you don't it will twist, damage the pipework supporting it, and before you know it you'll be very damp indeed.

10 As you open the nut water will start to drip out into your tray. There shouldn't be too much at this stage and you should easily be able to contain it in the bowl.

11 Now open the bleed valve again. As you do this the flow of water will increase considerably. When the bowl is almost full close the bleed valve, swap the full bowl for the empty one and open the bleed valve again whilst your mate empties the first bowl. Do this until the radiator is empty. Once empty make sure you fully close the bleed valve. Of course, if you have a water vacuum use this as it's much faster and easier – although be ready to close the bleed key in case the vacuum fills and stops sucking.

12 Undo the other radiator tail. There might be a few drips of water, but nothing too exciting.

13 Lift the radiator off its brackets and quickly turn it upside down. It can now be taken out of the house.

There's usually some water still in the radiator and it's usually filthy, so be careful when removing the radiator. If you do get any on your carpet vacuum it up ASAP and give the area a quick rub with a sponge and some warm soapy water.

14 Check that none of the radiator valves are leaking and that you've put caps on all the open valves.

REFITTING THE RADIATOR

Refitting the radiator after you've finished decorating is really just following the instructions for fitting it, but in reverse.

Tools and materials
- Pump pliers
- Large and small adjustable spanner
- Bleed key
- Tissue paper or bowl

1 Carefully turn the radiator the right way up, bearing in mind it almost certainly contains at least some utterly filthy water. Stuffing the radiator tails with tissue paper before you start is a good idea.

2 Set the radiator on its brackets, ensuring it's held in place top and bottom. Note that there are little plastic plugs on the brackets. They're not essential, but check that yours are in place as they have a habit of dropping off, never to be seen again.

3 Now push the radiator valves on to the tails and hand-tighten, then, whilst supporting the valve with a set of pump pliers, fully tighten the nut with an adjustable spanner by giving it an extra three-quarter turn.

4 Open the radiator valves one at a time, checking for leaks each time. For the lock-shield (the one you had to take the head off and turn with an adjustable spanner) read your notes and turn it back the right number of turns.

5 Bleed the radiator.

6 If you have a sealed system bring the pressure back up to 1bar via the filling loop.

Fitting a new radiator

When fitting a new radiator it's nigh on impossible to reuse the old brackets, but life is a lot easier if you reuse the old valves, *ie* leave the old ones on the pipework. The reason for this is simple; if you leave the old valves in place you don't have to drain and refill the CH system and there are no pipework alternations. However, you will need to pick a replacement radiator that's the same width or just slightly narrower.

If you removed a very old radiator the odds are it was an imperial size and you'll not find an exact replacement, but you can buy radiator 'extensions' that can make up for a gap of between 10-40mm on each side.

If you buy a replacement that's wider than the old one you usually have no choice other than to alter the pipework, and whilst you're there replace the valves – see 'Fitting new radiator valves'.

Tools and materials
- Adjustable spanner
- PTFE tape
- Radiator spanner
- Level
- Tape measure
- Hammer drill
- 8mm screws and rawlplugs, or spring toggles for plasterboard
- Screwdriver set or impact driver

1 Unpack the radiator and find the brackets – they're usually hidden inside the fins.

2 Fit the bleed valve to the top hole in the radiator at the end that's going to be easiest to get to once the radiator has been fitted. Use an adjustable spanner to tighten it, but don't overdo it or you'll cause the rubber seal to twist out of position – make sure the bleed valve is closed.

3 Fit the blanking plug to the other top hole in the radiator.

4 The bottom holes usually have some sort of plastic blanking plug in them to stop the thread getting damaged. They can sometimes be a real pain to get out and there's no ideal approach, so just go for them as you feel fit.

5 Take the radiator tail and hold it in your left hand. Take a roll of PTFE in your right hand. Take a bit of tape off the bottom of the roll and place it on top of the thread.

TIP *If you're using the old valves you might as well use the old tails. To remove the tails fit a radiator spanner inside the tail and turn. In the old days these always used to work but these days the internal size can vary so much you might want to opt for the adjustable version. Also, many tails can now be unscrewed with an adjustable spanner on the outside.*

Holding this bit in place take the roll around the thread clockwise for about 12 turns. Remember the important point that was made earlier in this book – you need to wrap the PTFE in line with the thread, *ie* clockwise, so that as you tighten the thread the tape is tightened and pushed down into it.

I've heard tell that four wraps of PTFE is sufficient, but personally I add at least twice this amount and I've known people treat the radiator tail as if it's an Egyptian mummy. Don't be daft about it, but err on the side of caution – you really don't want the radiator to leak from these tails as it can be a real job to rectify.

6 Fit the tails and tighten them up. Depending what type of tail you have you'll either need an adjustable spanner or a radiator spanner.

7 Site the radiator in position on the floor so that the tails and valves are in the correct position for your pipework. Mark on the wall the position of the brackets on the back of the radiator.

8 Move the radiator aside and draw a line down the wall using a level to mark the bracket position.

9 If you're fitting new radiator valves it's easier to hang the radiator and then adjust the pipework to match this position. If you are doing this you want to hang the radiator so that the base is 15cm above floor level – this allows the best flow of air around the radiator.

To get this height, fit a bracket to the radiator (put the little plastic inserts on to the bracket first) and measure the distance from the top of the bracket to the floor. Take this measurement, add 15cm and mark this total height on the wall. Use this as the position for the top of the bracket. Now jump to step 11.

10 If you're using existing radiator valves you need to get the radiator height just right. To do this, measure the height from the floor to the centre of the existing valves. In this example let's say it's 175mm. Then, with the radiator on the floor, measure the height of the radiator tail centres from the floor – we'll say this is 25mm.

TIP *Most of the time the radiator valves have a bit of movement in them allowing you to push the pipework down or lift it up a little. Check to see if this is the case, and if it is take your measurements about midway so you have room for adjustment if you get any measurements wrong.*

If one valve and its pipework has far less movement than the other use this side for all your measurements.

Always remember that you want the base of the radiator about 150mm above the floor.

Subtract the latter dimension from the former (175 – 25 = 150mm) and this is the height the radiator base needs to be.

Now, with the radiator resting on the floor, put a bracket on it (make sure the plastic inserts have been fitted) and measure the height from the top of the bracket to the floor. Now add on the height you got earlier and you have the height of the top of your bracket.

By the way, the little plastic inserts are there to cushion the radiator and stop it making any noise as it heats up and cools down.

11 Mark the drill hole positions on the wall. There are usually two round holes and two long slots. The holes offer better support but allow no margin for error. The slots allow you to adjust the height of the brackets but also mean the bracket can drop under the weight of the radiator. I tend to take the middle road and fit the first bracket using the holes, so it's firmly in place and can't move at all. Then I use the slots on the second bracket because that allows me to get the radiator exactly level.

An alternative approach is to fit the brackets using the slots to start off with and once the radiator is level and at the right height take the radiator back off and mark and drill the holes. This is much more time-consuming but is probably worth it if you're fitting a big, heavy radiator. A laser level is also a handy tool to have around when trying to get large radiators exactly level.

The brackets can be set so that the supporting arm is narrow or wide. Use the narrow setting for double-panel radiators and the wider setting for single-panel radiators.

12 Drill the holes for the first bracket and screw it back to the wall.

13 Set the second bracket by putting a level on top of both brackets and moving the second bracket until it's perfectly level with the first. Mark the slots for this second bracket and loosely secure it to the wall.

TIP *It's always best to use size 10 screws for radiators. If you're hanging the radiator on a plasterboard wall try to find a wooden stud if possible. If you're just hanging it from the plasterboard try using spring toggles to hold the bracket in place. If it's only plasterboard holding your brackets in place I'd also apply sealant – or one of those glues that suggest an absence of nails – to the back of the bracket as well as the spring toggles. This will spread the weight a bit more, although it will also make removing the brackets rather awkward in the future. All that aside, the best approach is to cut out the plasterboard, fit a batten where required and then put the plasterboard back. However, this is a tad time-consuming and as a result is rarely done.*

14 With the screws only half-tight so they just hold the bracket in place, get the two brackets exactly level with each other. Once there fully tighten the screws.

15 Fit the little plastic inserts.

16 Lift the radiator into position.

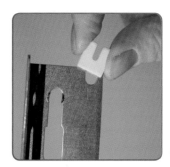

17 Fit the radiator tails to the radiator heads. Hold the head steady with a set of pump pliers and tighten the tail using an adjustable spanner.

18 Open the radiator valves and check for leaks. If all is dry open the bleed valve and bleed the radiator.

Fitting new radiator valves

The most common reason for fitting new radiator valves is so that you can replace the traditional valves with thermostatic ones (TRVs). These are a very good way of controlling how warm each room in your home gets. They're especially handy if you prefer your bedroom comfortably cool when the rest of the house is warm.

One up from the TRV is the programmable radiator valve. These allow you to control the heat output from a radiator based on the time of day and the day of the week. These can be very useful in offices and schools but are also handy if you have one room in your home set aside as an office.

Tools and materials
- Bleed key
- Pump pliers
- Adjustable spanner
- Old towels

1 Open the radiator valves you wish to change and drain the CH system.

TIP *On large systems you might wish to freeze the pipework rather than drain down the entire system. There are two approaches to pipe freezing: the first uses a spray can to fill a foam pipe cover with a freezing liquid; the second is plugged into the electricity and runs the same way as a fridge. Personally I'd always opt for the second approach; they're readily available for hire, cheaper to run and much more reliable to use.*

2 To ensure the radiator is empty open the bleed valve. You should wait until the flow from the hose draining the system has slowed down before doing this. As you open the bleed valve you should hear a hiss as air runs into the radiator.

3 When the flow has stopped completely hold the valve head steady with a set of plump pliers and open the radiator tail with an adjustable spanner. There will always be some dirty water inside, so lay some old towels underneath to catch it.

4 Most radiator valves now come with ½in tails, whereas most old radiators have ¾in tails. If this is the scenario you're faced with you'll have to change the tails as well as the heads – see steps 5–6 in 'Fitting a new radiator'.

If the old tails are the right size and seem to be in decent condition, then you're best off leaving them in place and reusing them.

5 Take the old valve off the pipework by holding it in place with a set of pump pliers and undoing the nut at the base with an adjustable spanner.

6 It's possible to reuse the nut and olive and just fit the new valve on top. However, it's better to replace the nut and olive as well. For information on how to do this read 'Removing compression fittings' (Chapter 6).

7 Fit the new valve, hand-tightening the nut at the base and then attaching the radiator tail. Once the valve is fully in place hold it steady with a set of pump pliers and tighten the base with an adjustable spanner. Now do the same with the nut on the tail. Be aware that pump pliers can damage the chrome on the valve so you might want to protect the valve by putting a wad of cloth between it and the pliers.

8 If you're fitting TRVs you need to fit the head now. This varies from valve to valve but generally you need to set it to the highest setting (usually 5) and push it on to the base. Now tighten the securing ring at the base of the head.

9 Close all the valves. Check everything is tight and refill your CH system.

10 If all is dry, open the new radiator valves and bleed the radiators. If you have a sealed system recharge it via the filling loop.

 TIP *As a general rule, don't fit TRVs to all the radiators in your home – always leave at least one with a traditional valve, and leave this radiator on all the time. The reason for this is so that there's always an open circuit in the CH system. The one most often chosen for the traditional valve is the bathroom radiator or heated towel rail.*

Fitting a heated towel rail

A heated towel rail is really just a decorative radiator. However, the design and colour of a heated towel rail means that they usually only deliver a fraction of the heat of a radiator of comparable size.

So before you do anything, measure the size of the room that the heated towel rail is going in and then go online and find a 'heat requirement calculator'. Lots of websites offer some form of calculator and all you need to do it enter the room dimensions and fill in some details about the room, such as letting the program know if you have double glazing or if it's north facing etc. This will advise you what kilowatt (kW) or British thermal unit (Btu) output you need from your radiator. Bear in mind that this is a minimum value, so opt for a rail that gives a slightly higher output – especially if you're going to drape the rail in towels.

Read the earlier sections in this chapter on 'Fitting new radiator valves' and 'Fitting a new radiator' for general fitting advice, but also bear the following in mind:

- Chrome heated towel rails produce far less heat than an equivalent-sized white rail.
- A heated towel rail is usually set slightly higher from the floor than the 15cm of a traditional radiator. This is for aesthetic rather than heating reasons.
- You generally use straight radiator valves for a heated towel rail. This is because the valves attach underneath the rail as opposed to the traditional angled radiator valves that fit on the side of a radiator. If the pipework is coming out of the wall you could stick with angled valves.
- You usually have to adjust the CH pipework to fit a heated towel rail. If this is the case it's much easier to fit the towel rail and get this perfectly level before fitting the pipework to the rail.

Common central-heating problems

NOISY CENTRAL-HEATING SYSTEMS

The most common cause of noise in a new central-heating system is the pipework expanding and contracting as it heats up and cools down. In this instance the noise tends to be an extremely irritating series of high-pitched 'dink, dink, dink' noises. There are a number of causes of this:

- The pipework is rubbing against other pipes.
- The pipes pass through notches or holes in joists and floorboards that are just too small, so the pipework rubs against the wood.
- The radiators are making the noise as they move on their brackets.

Finding exactly where the noise is coming from can be a challenge in itself and I'm afraid you're on your own there. However, having found the source you can cure it by either slightly widening the holes or notches or by putting foam insulation around the pipework to absorb the movement.

If it is the radiators that are making the noisy try putting plastic inserts on top of the brackets – you'll need to remove the radiator first (see above for more details).

TIP *The saying 'a stitch in time, saves nine' has a bearing here, as over time pipework rubbing against a joist will slowly wear away until a leak develops.*

On older central-heating systems the usual cause of a sudden noise in a previously quiet system is a blockage of some sort in the boiler or pipework. This is often called 'kettling' and can generate the most horrendous bangs and booms.

The usual cause of the blockage is limescale build-up. The blockage doesn't have to be total, just a small restriction can be enough to cause localised temperature and pressure changes, which in turn cause steam to be released. It's this steam that makes the noise.

The easiest way to cure it is to add 'noise reducer' to the system. A number of manufacturers product this and it's simply a scale remover that you pour into your CH system exactly the same way as you would inhibitor – see 'Adding inhibitor to the system' earlier in this chapter.

This will usually start to reduce noise levels within a few hours and will cure it completely within a few weeks – although it often gets worse before it gets better. If it doesn't then you might have a physical blockage in the pipework, *eg* a valve is only half open or something has worked loose from inside the boiler and is now blocking the pipework. If this is the case you're probably best off calling in the professionals, although it's always worth adding a further bottle of 'noise reducer' first just in case there was a lot of scale in the system.

SOME OF MY RADIATORS WON'T GET HOT

This is a common problem when more radiators have been added to an existing system or in old systems where the radiators just get cooler and cooler over time. In the former case the cure is a process called 'balancing the system'.

Balancing your central-heating system

The water within the CH system will always try to take the easiest route back to the boiler and if it can do this without going through umpteen radiators it will. With an 'unbalanced' system what normally happens is that some radiators get very hot very quickly, whilst others remains cool or just plain cold. To bring 'balance' to the system follow these steps:

1 Remove the lock-shield from those radiators that are always cool and use a small spanner to turn the head anticlockwise until the valve is fully open. A lockshield is the valve on the radiator that doesn't have a head that you can turn. To adjust it you need

to first remove the cap (or shield) either by unscrewing it or just pulling the head off. Once the shield is off you can open and close the valve by turning the head using a small adjustable spanner.

2 Turn your CH system on and note which radiators get hot first.

3 Remove the lock-shield cap from these radiators and turn the head clockwise a quarter turn. What you're trying to do is create more resistance to flow within those radiators that are getting plenty of hot water in the hope that this will encourage the water to go into the radiators that are still cold.

4 Repeat steps 2 and 3 for all those radiators that get fully hot, returning to the first hot radiators and giving them an additional quarter turn clockwise if necessary.

5 Turn the system off again, let everything cool down and repeat steps 2–4 until all the radiators are getting an equal amount of heat. This process can take a while to get right, but it has the virtue of being very easy to do.

Pump speed and size

If you've added more radiators to an existing system you might find

that you need a slightly more powerful pump. However, before you go out and buy one check the speed setting of the current pump. Most come with three speed settings and a little switch on the case to change from one to another. If your pump is currently set to 1 or 2 then move it up a notch. If it's already on 3 then you might need a more powerful pump.

The usual domestic pump is called a '15/50', the more powerful version is called a '15/60' and will fit in exactly the same location as the old pump without the need for pipework alterations – see 'Fitting a new central-heating pump' below.

Powerflushing

If your radiators have slowly got colder and colder over the years the odds are there's a lot of gunk in the system. In this instance your best course of action is a powerflush, and maybe replace some of the really bad radiators – see 'Powerflushing the system', 'Removing a radiator' and 'Fitting a new radiator' earlier in this chapter.

MY DOWNSTAIRS RADIATORS ARE COLD BUT UPSTAIRS IS FINE

The usual cause of this is that the pump isn't working. To check this out turn the CH off and put your hand on the pump – be careful, as the pump can get very hot. Now get someone to turn the CH on. Within a few seconds you should be able to feel the pump vibrating slightly under your hand. If it doesn't there are two probable causes:

The pump is stuck

It's not unknown for the impeller to get stuck. To free it follow these steps:

1 With the CH off, put a sponge under the pump and remove the central screw. A small amount of water will emerge.

2 Insert a fairly small flat-headed screwdriver into the hole and try to turn the spindle by hand. It should turn with little or no effort. If it doesn't, apply a little force to see if you can free it. Once it's turning freely put the screw cap back on.

The pump is broken

If the steps above didn't fix the problem and the boiler itself is firing up OK then the odds are the pump is broken. However, it pays to check that the pump is getting electricity and there isn't just a loose wire in the wiring centre.

The easiest way to test this is to purchase an electrical detector. Test it's working by applying it to a known live (don't apply it directly, just place it on the insulated wire leading to the known live) and then check the wire leading to the pump. If it lights up but the pump fails to start, follow the steps in 'Replacing the central-heating pump' below. If it doesn't light up, ask an electrician or heating engineer to check the wiring.

I'M GETTING NO HOT WATER BUT I HAVE CENTRAL HEATING

If you have a combi boiler then it's usually the flow detector, the diverter valve or the PCB in the boiler that's gone. All should only be replaced by a Gas Safe registered engineer.

If you have a heat-only boiler then this problem is usually caused by one or more of the motorised valves failing to open properly and will usually result in the boiler failing to fire up when

Courtesy of Bristan

on the 'hot water only' setting. See the section on 'Replacing the motorised valve' below to change this valve, and in the interim use the immersion heater to generate your hot water. A common giveaway for this scenario is that you do get hot water if you turn the programmer to hot water and CH.

If you have a direct hot-water cylinder, *ie* it's powered by an electric immersion heater rather than the boiler, you need to read the section on 'Replacing the immersion heater' below.

I'M GETTING HOT WATER BUT THE RADIATORS WON'T GET WARM

Check that the pump is working. If is it then it's probably the motorised valve playing up, so read the section on 'Replacing the motorised valve' below.

THE RADIATOR VALVE IS LEAKING

The most common time for radiator valves to leak is when you've just balanced the system (see above) and the valves have been adjusted for the first times in years. The leak will usually appear at the top of the valve through the gland nut, and in most instances you can stop it by giving this nut a clockwise tweak.

If this doesn't work you may have to repack the gland nut. The process for this is exactly the same as for a tap – see 'Repacking the gland nut' (Chapter 7). But there's a caveat to this advice; some radiator valves use a rubber washer inside them, and attempting to repack these can cause the spindle of the valve to fly out and spray dirty CH water all over the place. With this in mind I'd advise you to drain the CH system first before playing around with the valves and, having gone to this trouble, seriously consider changing the entire valve.

MY THERMOSTATIC RADIATOR VALVES HAVE STOPPED WORKING

There are two parts to a thermostatic valve, the head and the main body. The head can usually be removed by undoing a nut at its base – although some also require you to remove a grub screw. Once removed you'll see a pin poking up from the middle of the valve.

In this make the pin is a bit chunky

They are often far more pin-like

Over time this pin tends to get stuck in place and as a result the valve stops working. To rectify this is often as simple as just tapping the pin lightly with a hammer so that it starts to move freely again.

MY BOILER HAS LOST PRESSURE

If you have a combi boiler or a high-pressure CH system (see Chapter 2 for more details) you might find that your boiler suddenly stops working, and when you look at the pressure gauge you'll notice that the pressure has dropped below 1bar. There are a number of causes of this:

You forgot to recharge the system

Having bled one or more radiators you didn't recharge the system via the filling loop. If you think this might have been the case just recharge it now and all should be well – see 'Using a filling loop' in Chapter 2 for more details.

There's a problem with the expansion vessel

In a high-pressure system (also called a sealed system) the pressure in the central-heating pipework and boiler will rise as the water heats up. To stop the pressure rising too high the system comes with an expansion vessel. This is a sealed container with a rubber membrane inside. On one side of this membrane is air under pressure and on the other side is the CH water. As the CH water heats up it pushes harder against the membrane, which gives ground and in doing so gives the hot water more room. This in turn lowers the pressure of the hot water and all is well with the world.

So what can go wrong? Well, first off the air in the vessel can lose its pressure. When this happens there's nothing for the water to push against so the vessel fills with water even when that water's cold. This leaves no room for any expansion when the water gets hot and the system starts to over-pressurise.

The second common failing is that the membrane tears and the vessel starts to fill with water, and so once again you lose any room for expansion when the water's hot.

Rather than let the pressure get

CH water cold   CH water hot

dangerously high your CH system comes with a second safety device called a pressure relief valve. When the pressure gets too high (usually over 3bar) this opens up and lets the water escape. If your boiler was installed correctly this escaping water should end up running down your outside wall in such a fashion as to catch the eye and warn you that something's wrong.

Sadly, once this pressure relief valve has opened up it has a habit of not reseating itself properly and continuing to leak water despite the pressure now being normal.

So having seen water escaping from the pipe and discovered that the boiler pressure is too low, what should you do? Well, even though it's quite an easy fix what you have to do is call a Gas Safe registered engineer who'll either recharge the expansion vessel or replace it. They might also play around with the pressure relief valve and, if they can't get it to close properly, replace this as well.

There's a temptation to try to fix all this yourself, but please fight it, as these are very important safety devices. At best you might invalidate any warranties, at worst you might blow up your house, which is always embarrassing.

Replacing the central-heating pump

This is effectively the heart of the CH system. Because everything moves through this pump it tends to suffer more when the system hasn't been maintained properly and is one of the more likely items to fail. The pump is also one of the prime areas where you can save a bit of money on your heating – not a huge amount, but a saving nonetheless.

The traditional pump is set to a single speed, so when the system starts up and heat needs to be distributed as fast as possible the pump is set too low. When the system is up to temperature and we want it to just tick along it's set way too high. So aside from a brief few minutes a traditional pump rarely pumps at the optimum rate.

To get around this the EU has insisted that pumps become a little smarter and a little more economical, resulting in a line of vari-speed 'smart' pumps. Not only should these deliver just the right amount of oomph to your system exactly when it needs it, but in doing so it should save you enough money to pay for itself in just a few short years.

If you're going to fit a vari-speed 'smart' pump you ought to clean the system first. These pumps tend to work by measuring resistance to flow, so if your system is thick with gunk they tend to assume that most of the thermostatic radiator valves are shutting down and that therefore the system is already up to temperature. As a consequence the pump slows down. The technology is changing all the time so no doubt they'll

get around this particular issue but for now it makes sense to clean the system before fitting a new pump.

Replacing a pump can be a messy task if the CH water is dirty, so put a few old towels under the pump before you start.

1 Use a small adjustable spanner to close the pump isolation valves by turning the heads clockwise.

2 The old pump should have a large arrow stamped on it. Note down which way this is pointing so that you fit the new pump in the same direction.

3 Steady the head of the isolation valve with one set of pump pliers and use another set to turn the nut anticlockwise if looking at it from the pipework to the pump.

Dirty water will escape at this point so set a bowl or some old towels underneath the pump.

Just crack the nut to start off with, and wait for the flow to abate. If it just seems to keep on coming then either you haven't fully closed both isolation valves, or one of both of the valves are faulty.

These nuts can become corroded into place and turning them can be a real challenge. If you're having no joy try heating the nut with a blowtorch or shocking it with a freezing product such as 'Crack-it', and then have another go at it.

4 Once the two isolation valves are open the pump will just slip out of position.

5 You might want to ask a qualified electrician to do this step. Turn off the electricity to the pump – always take the fuse out and put it in your pocket so that the electricity can't be turned on again without you knowing about it. Open the cover of the pump and remove the three wires (live, neutral and earth).

6 Remove the cover of the new pump and fit the three wires into the right connections. This is usually achieved by pulling down on little levers; this opens the holes and lets you push the wires in. Refit the cover.

7 All domestic pumps fit into the same gap, so the isolation valves should already be in the right position. However, the washers have probably seen better days so fit the new ones that usually come with the pump.

8 Slide the pump into position, making sure that the arrow is pointing in the right direction. In an ideal world the pump should be on a vertical pipe and pumping upwards.

If it's on a horizontal pipe the head should be slightly higher than the plane of the pipework but not vertically above nor under it.

TIP *If you're fitting a pump into an entirely new position and not just replacing the old one, bear the following in mind:*

- *The vent pipe and feed pipe on an unvented CH system should be just before the pump. They should all be fairly close together on the flow pipe in the order vent, feed, pump.*
- *The motorised valves should be after the pump.*
- *Most systems need a bypass valve to allow water to still flow if the motorised valve is closed. This is fitted between the pump and the valves and leads to the main return pipe.*
- *You should try to avoid elbows close to the pump – use machine bends if possible.*
- *Make sure the pump head is accessible.*
- *Make sure the pump isn't touching anything.*
- *Avoid placing the pump low in the system where debris can settle within it.*

9 Tighten the isolation valve nuts.

10 Open the valves and check for leaks.

TIP *The following scenario isn't unknown (ie it happened to me when I first started plumbing). You're happily fitting your new pump when you knock against the pipework and turn to discover that the seemingly impossible has happened – the*

nut has fallen off the isolation valve. You try to put it back on, it just won't happen, and you start to panic...

Well calm down. The nut has two slots cut into its side. Align these so they're uppermost, push the nut back on to the valve, and relax.

11 Open the screw cap in the middle of the pump to let water enter the pump.

12 Turn the electricity and your CH back on.

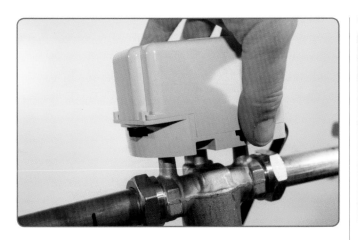

Replacing the motorised valve

There are actually very few moving parts in a CH system. Sadly the motorised valve is one of them, and as such is often a cause of the system failing.

There are two types of motorised valve: two-port and three-port, often called a 'mid position' valve. It's easy to tell the difference as the ports they refer to are actually pipes; so count the pipes going in and out of the valve and you'll know which one you have. In plumbing parlance the two-port valve is part of an S-plan CH system, whereas the three-port valve lies at the heart of the Y-plan system.

Most motorised valves let you remove the 'motor' part – the actuator – without having to change the main body of the valve itself. This is handy, as it's often the motor that burns out before the rest of the valve.

The way you remove the head depends on the manufacturer. Some, usually the ones with a plastic cover, have a little button on the side that you just press and then pull the plastic cover off. Others have metal clips at the base of the actuator and these you can just pull off. The rest, usually those with metal covers, require you to remove the cover first and then unscrew the actuator from the body of the valve – but first make sure the electricity to it is turned off.

TIP *The actuator has more than just live, neutral and earth connections. With this in mind it's best to always buy the same make and model and to rewire them by turning off the electricity and cutting off the cable of the old actuator just before the wiring centre. You can then fit the new one and just replace each wire separately.*

If the photo puts you off at all – and let's face it, it should – call out an electrician.

The reason the motor burns out is often due to the valve mechanism slowly filling with debris so that turning becomes harder and harder. To avoid the same thing happening to your new actuator put a small adjustable spanner on the valve head – I'm assuming you've already removed the old actuator – and turn it a few times.

The head only turns a tiny fraction, between twelve o'clock and two o'clock at most, so don't try to wrench it any further. If it turns freely within this small arc then all should be well. If it's very stiff to turn then the new actuator is just going to burn out like the last one, so you're best off replacing the entire valve.

To do this, drain the CH system. For two-port valves note the direction of the arrow on the existing valve and ensure you put the new one in the same way. Some manufacturers seem to find stamping an arrow on their valves to be a real challenge and have opted to replace this exceptionally clear indication of direction with a more enigmatic 'A' and 'B', where flow is from A to B. No, I can't understand why anyone would want to do this either.

If you have a three-port valve the ports are labelled 'A', 'B' and 'AB'. 'A' goes to the radiators, 'B' goes to the hot water and 'AB' is the in pipe coming directly from the boiler via the pump.

Make sure the valve head is accessible. It's always best to fit this above the main body of the valve so there's no risk of water dripping into the mechanism.

Note the little lever on the side of the motorised valve. This allows you to manually open it by pushing the bar towards the 'MAN' side and then pushing it down so that it locks in the 'open' position. This is very handy if the valve breaks, as it means you can keep your hot water and CH running whilst you find a replacement. On actuators made from plastic this lever is usually replaced by a black plastic bar, but the principle remains the same.

Fitting a new hot-water cylinder

There are four types of hot-water cylinder:

UNVENTED HOT-WATER CYLINDERS

If you have a fairly modern home the odds are you have one of these. You can usually identify them by the fact that they're have a steel skin and lots of valves and pipes. Under no circumstances should you try to replace or repair one of these by yourself.

The water in an unvented cylinder is under pressure. The problem with hot water under pressure is that you have to be very careful to keep both the pressure and the temperature under control. If the temperature rises above 100°C and the pressure is above 1bar you've created the perfect conditions for a bomb. And let me make it clear, this is not a little 'fizz, squeak' bomb; this is a 'where did my house go?' bomb.

To stop this ever happening the unvented cylinder comes with an array of safety devices which must be fitted and serviced by a suitably qualified 'competent person', *ie* he has an ID card to say he's fit to work on unvented hot-water systems.

THERMAL STORE

This is effectively the reverse of a normal hot-water cylinder, in that its main body is filled with hot CH water direct from the boiler and the coil inside it is where the domestic hot water is generated. We're not going to look at these in depth but here are a few tips and notes:

- You can identify a thermal store by virtue of the fact that you don't have a large storage tank up in the loft and your hot water is at mains pressure.
- They're often made from copper rather than steel.
- You don't see many of them about, as the unvented cylinder works much better and lasts far longer.
- The main problem with them is that the coil inside gets caked with limescale and, as a result, the water doesn't get as hot as it passes through. They also suffer in winter when the mains cold water is close to freezing and needs far more energy to bring it up to the desired temperature.
- On the plus side you're not going to run out of hot water with a thermal store, provided your boiler is working properly, so they're very handy where there's a large demand for hot water.
- If you have a thermal store it makes a lot of sense to fit a water softener in your home to extend the life of the system – see Chapter 10.
- If your thermal store is reaching the end of its life I'd think about replacing it with an unvented hot-water cylinder.

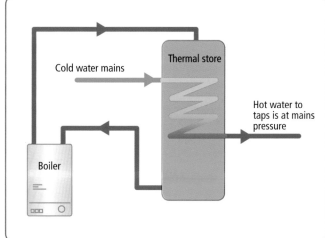

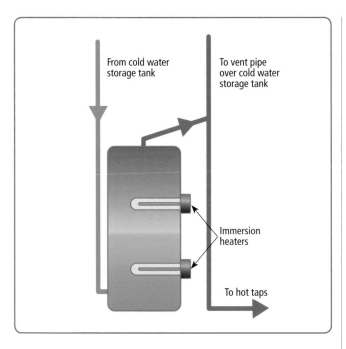

From cold water storage tank

To vent pipe over cold water storage tank

Immersion heaters

To hot taps

Hot water pipework

Cold water storage tank

Cylinder

Boiler

Heating coil

DIRECT VENTED HOT-WATER CYLINDER

This looks exactly like a standard hot-water cylinder except that it's not connected to your CH system. Instead the water is heated by one or more electric immersion heaters. You can quickly identify these by the lack of pipes – just one pipe coming in at the bottom, and another leaving at the top.

This set-up is more usually found in rural areas where the CH is provided by oil or electricity, although, as the price difference between electricity and other forms of energy diminish, they may well become more popular.

The most common reason for replacing these cylinders is when the immersion heater fails and you can't get it out of the cylinder without tearing the thin copper walls of the cylinder itself.

To replace them just read the instructions for an indirect cylinder below.

INDIRECT VENTED HOT-WATER CYLINDER

Most UK homes over 20 years old have one of these. They work by filling the body of the cylinder with cold water supplied from a water storage tank above. This is then heated by passing hot CH water around a coil inside the cylinder. In other words the boiler heats your water indirectly, hence the name.

Most cylinders will also have an electric immersion heater built into the top as a standby in case the boiler breaks down.

The great thing about these cylinders is that they can last for decades without anything ever going wrong with them, for the simple reason that there's very little in them to go wrong. However, eventually the copper walls start to leak, usually at the base.

To replace a hot-water cylinder, follow these steps:

Tools and materials

- ■ Small spanner
- ■ Hosepipe
- ■ Jubilee clip for hose
- ■ Pump pliers
- ■ Adjustable spanner
- ■ 22mm pipe cutter
- ■ Set of screwdrivers
- ■ Wire cutters
- ■ PTFE
- ■ Jointing compound suitable for drinking water
- ■ Utility knife
- ■ Timber for base of cylinder
- ■ Drill and hole-cutter

1 Turn off the electrical supply to the immersion heater(s) and, for an indirect cylinder, turn off the CH heating system.

> **TIP** *Don't just turn off the electricity supply – also remove the fuses and put them in your pocket. Many an electrician has been fried by someone walking by and flicking the switch back on.*

2 Turn off the hot water by closing the tap connecting the cylinder to the cold-water storage tank. If this tap doesn't close turn off the mains cold water, either at the stop tap or at the isolation valve feeding the storage tank. Read Chapter 2 for guidance on how to identify this tap.

3 Open all the hot-water taps. If you can close the tap connecting the cylinder to the cold-water tank the hot water should stop running after 10–20 seconds. If you've had to turn off the cold water altogether then it will take at least five to ten minutes as the tank in the loft drains down.

4 The hot taps should now have stopped running. However, this only means that you've drained the water down to the top of the cylinder; the cylinder itself is still completely full. Near the cylinder there's a drain cock – sometimes on the cylinder itself, but usually at the bottom of the cold-water feed pipe to the cylinder.

Attach a hose to this, using a jubilee clip to get a watertight seal. Run it into a waste that's lower than the base of the cylinder and then open the drain cock and drain the cylinder down.

If the drain cock doesn't open try tightening it up really tight and opening it again. If this works you'll need to re-washer the drain cock before you fill the new cylinder.

Still no joy? Then undo the pipe at the top of the cylinder or remove the immersion heater – only do this if you have an immersion heater that fits into the top of the

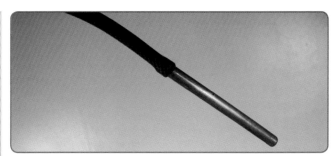

cylinder (read 'Replacing an immersion heater' on page 197 for more details). Even then expect a small amount of water to come out. Now fit a short length of copper pipe to the end of your hosepipe and put the hose into the cylinder so that it rests on the base.

Take the other end of the hose to a waste point lower than the base of the cylinder and suck on it to get the siphonic action going. Try not to get a mouthful of water, as it's usually very dirty at the base of the cylinder. Let the cylinder drain completely.

If you have a direct cylinder jump to step 7.

5 On an indirect cylinder you'll notice two pipes going into its side. These are part of the CH system and connect to the coil inside. They're usually referred to as the flow and return pipes and before you can remove them you need to drain down the CH system. To do this turn off the water supply to the system (see Chapter 2) and open a drain cock on one of the radiators downstairs.

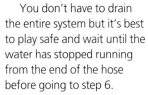

You don't have to drain the entire system but it's best to play safe and wait until the water has stopped running from the end of the hose before going to step 6.

6 Use a set of pump pliers to undo the nuts holding the CH heating pipes in the cylinder. It's important that these pipes don't get confused, *ie* the upper most pipe must go into the uppermost hole of the new cylinder, so identify them in some way.

7 Use pump pliers to undo the nut at the top of the cylinder. Sometimes this can be very difficult. If this is the case use a pipe cutter to cut through a nice, clean, straight piece of the pipe that emerges from the top of the cylinder.

8 Undo the nut at the base of the isolation valve or gate valve that's feeding cold water to the cylinder. If you had to drain the cold water completely because this valve failed to close then this is the time to remove this old valve and replace it. If you didn't have to drain the cold water completely remember that this is the valve keeping you dry, so whatever you do don't undo the top of it.

9 You might want to call in an

electrician at this stage. Make absolutely sure that the electricity supply to the immersion heater has been turned off. Undo the small nut at the top of the immersion heater and remove the cap. Now undo or cut the three wires to the immersion.

10 The indirect cylinder should have a cylinder thermostat attached to it. This is usually held in place by a wire. Undo the wire and remove the thermostat. If you pull aside the wire you can simply pull the thermostat out of the foam into which it's usually inserted.

11 Make sure the drain cock you opened earlier is closed. Stuff all the open holes in the cylinder with tissue paper and carefully remove the cylinder from the house. Be aware that it will almost certainly contain some very dirty water.

TIP *Old copper cylinders are well worth taking down to the local scrapyard. The money you get for it will go a long way towards paying for the replacement cylinder.*

12 Life is much easier if you're replacing the old cylinder with one that's roughly the same size. Either way the following points need to be borne in mind:

- The base of the cylinder must be fully supported, not just resting on a few joists. Don't use chipboard or MDF as these weaken when wet; always use floorboards or 19mm plywood.
- Remember that the average cylinder will weigh about 145kg when full, or just under 23 stones. If you don't think two people could stand on the base you've just made, make a stronger one.

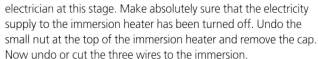

- The base of the cylinder is bowl-shaped to give it strength, which creates an air gap underneath it. The air in this space expands and contracts as the cylinder heats up and cools down and this can cause the base to weaken. To stop this happening always drill a small hole in the base, directly under the cylinder, so that the air gap is connected to the room in general.

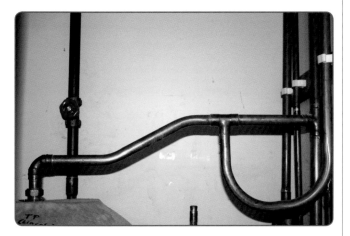

This pipe also includes a gravity loop for a mixer shower

- The pipe coming from the top of the cylinder is the main feed pipe for all your hot water. This must always rise gently so that the air released when the water is heated can vent away. It should end in a tee; the pipe going up is the vent pipe, which should terminate above the cold-water tank – see Chapter 2 for more details. Under no circumstances should there be a valve or any other potential blockage between the pipe leaving the cylinder and the end of this vent pipe.
- The electrical flex connecting the immersion heater(s) to the cylinder must be heatproof and the supply to the immersion heaters must come directly from the main consumer unit – in other words get an electrician in to fit and test it.
- On an indirect cylinder the thermostat should be positioned about one-third of the way up the cylinder, between the flow and return pipework. It must be in contact with the copper surface of the cylinder, so use a short-bladed knife to cut out a rectangle of foam insulation. Make sure you can see clean

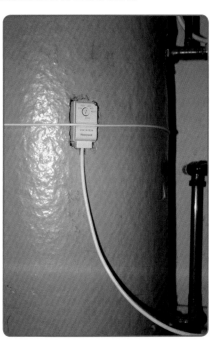

copper underneath before fitting the thermostat, using the wire provided to keep it in place – you might need to shorten this wire with wire clippers. The cut-out temperature for this thermostat is usually factory-set to 60°C. Just check what temperature the little arrow is pointing to – it should never be below 60°C, and the recommended temperature is 65°C.
- It's important that the two pipes going into an indirect cylinder

(the flow and return pipes) are the same way round as before, *ie* the uppermost pipe must remain at the top when the new cylinder is fitted.
- The cylinder fittings all require PTFE and jointing compound in order to ensure a watertight connection. They're often a nightmare to adjust afterwards, so make sure you apply lots of PTFE correctly, *ie* wind it clockwise around the thread and use jointing compound as well. Note that when you buy a new cylinder it doesn't come with these fittings so make sure you order them at the same time.
- When tightening up the fittings remember that the new cylinder is made from very thin copper, which will rip apart if you're too enthusiastic.

- See the section on 'Fitting a new immersion heater' below for details on how to install your new immersion heater.
- If you're using plastic push-fit fittings to connect up the new tank ask an electrician to test all the earthing around the tank afterwards.
- You need to ensure that the new cylinder can be drained in an emergency, so make sure there's a working drain cock close to the base of the cylinder. If you have doubts about the existing drain cock just buy a new one

and fit the new mechanism into the old drain cock body – this works 99% of the time.

13 By adhering to the points listed above you should be able to get your new cylinder into place without too much trouble. Next you need to fill it.

Firstly check every joint and connector in the new cylinder and make sure they're all tight. Open all the hot-water taps.

If you just isolated the cold water at the tap next to the cylinder open this, otherwise turn the cold water back on and let the cold-water storage tank start to fill.

The cylinder will make all sorts of strange and worrying noises as it fills. Ignore these and check the cylinder for leaks. Most leaks can be fixed by just tightening the fitting, but some will require you to turn the water off, drain the cylinder and start again.

Once the cylinder is full water will start to come out of the hot-water taps. Starting from downstairs, make sure water is coming out of each tap and then turn it off. Repeat until all the taps are closed. If you get an airlock read 'Dealing with an airlock' in Chapter 8.

For an indirect cylinder you now need to refill and bleed the central-heating system.

14 Turn on the CH and set the programmer to hot water, or turn the immersion heater on. Once the water in the tank is hot check it all again for leaks – hot water is far more liable to leak than cold.

Replacing an immersion heater

The hot-water cylinder is, in essence, a very large kettle, and the immersion heater is the kettle element. Since the cylinder itself is nothing more than a container, there's very little that can go wrong. Sadly, the same is not true of the immersion heater, and you'll probably have to replace this at least twice during the lifetime of the cylinder.

If you have an indirect cylinder then the immersion heater is only used as an emergency backup, or possibly as a cheaper option in the summer months. If you have a direct cylinder then you'll have to replace the immersion heater as soon as possible.

Whilst fitting an immersion is fine you might wish to ring an electrician to get it wired up afterwards.

Tools and materials
- Electrical tester
- Small adjustable spanner
- Electricians' screwdriver
- Sponge/towels
- Immersion spanner
- Hammer
- Abrasive strip
- Jointing compound suitable for drinking water

TIP *If you have an unvented cylinder don't play around with the immersion heater. Firstly it is often set into the side of the cylinder and secondly it is also a safety device in that it will contain a thermal cut-out. As such you ought to call in a G3 qualified technician.*

1 Use an electrical tester to check that the immersion heater is actually getting power. If you think it isn't call an electrician.

2 If you're sure it's the actual immersion heater that's gone – and it usually is – follow steps 1–3 in 'Fitting a new hot-water cylinder'.

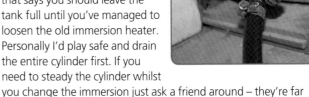

3 If your immersion heater is set very close to the top of the cylinder you can jump to step 3. However, some immersions are set into the side of the cylinder and you'll have to drain the cylinder before you can get at them.

To do this attach a hose to the drain cock on the pipework close to the base of the cylinder and drain the water away into a suitable waste pipe or outside gully.

There's a school of thought that says you should leave the tank full until you've managed to loosen the old immersion heater. Personally I'd play safe and drain the entire cylinder first. If you need to steady the cylinder whilst you change the immersion just ask a friend around – they're far less likely to flood your home than a full cylinder is.

4 Use an adjustable spanner to remove the bolt holding down the immersion heater's lid.

5 Double-check that the electricity is off and then remove the three wires (live, neutral and earth) inside the old immersion. Insulate the bare ends and tuck the cable safely out of the way.

TIP *Quite often it's the thermostat within the immersion heater that's gone. These days most stats are 'non-resettable', ie if they overheat or malfunction they don't reset themselves. Instead you have to press a little button set into the top, for which you'll need something like a biro or a pin.*

If, having reset the thermostat, the immersion heater starts working again you should get it checked by an electrician to see why it failed in the first place. Alternatively, just buy a new thermostat probe. To replace these you don't need to drain down anything. Just undo the two wires going into the old one, take it out and fit the new one. All very easy peasy – provided you remember to turn the electricity off first! If you have any doubts ask an electrician to do it for you.

6 When you remove the immersion you'll still release a small amount of water so put down some sponges and old towels.

7 You now need an immersion spanner or a very large set of pump pliers. You can buy a thin type of spanner but these tend to slip off all the time, so it's best to buy the thicker, box-spanner

style. Put this over the head of the immersion and try to turn the spanner anticlockwise.

If luck's on your side it will turn nice and easy. Sadly, this rarely happens, so your best bet is to take up a hammer and give the spanner a series of short, sharp taps to 'shock' the immersion into turning. If this doesn't work try applying some penetrating oil and letting it soak in for an hour or so, or try freeze shocking it using a proprietary spray.

The main thing to remember is that the cylinder's copper skin is very thin. If you apply too must pressure to the immersion spanner there's a very real danger of just tearing the cylinder open – at which point you need to buy a new one. There are times when this is the only option, but it's an expensive option well worth trying to avoid. Using the hammer allows you to apply a lot of pressure but reduces the risk of the cylinder tearing.

8 Once you've broken the seal the immersion heater can be wound out. There should only be a dribble of water but get ready with the towels anyway.

9 Check the edge of the hole where the new immersion will go and clean it with an abrasive strip if needed. It's the flat top to this hole that should be creating the waterproof seal, but this is often covered in foam – as you can see from the photograph. This needs to be removed by scraping it away with a utility knife.

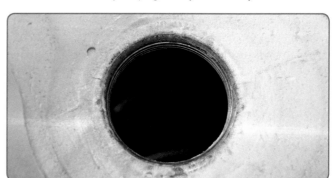

10 Put the fibre washer on the new immersion. This should be all you need to do to get a waterproof seal. However, it's not unknown for them to still leak so you might want to apply a small amount of jointing compound to the thread of the immersion heater, making sure it's suitable for potable water.

11 Tighten the new immersion into place.

12 Open the valve you initially closed to isolate the cylinder from the water supply. Leave all the hot-water taps open and let water start to emerge before turning them off. Keep an eye out for any leakage throughout this process. If you have a leak give the immersion a tweak with your immersion spanner. If this doesn't work you'll have to start again, applying more PTFE and/or jointing compound to the thread and checking the state of the washer.

13 With everything dry, reconnect the live, neutral and earth wires or ask an electrician to do this for you.

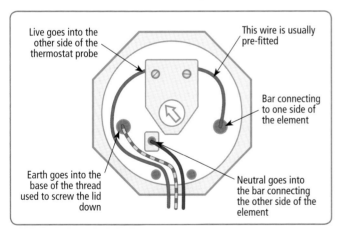

Live goes into the other side of the thermostat probe

This wire is usually pre-fitted

Bar connecting to one side of the element

Earth goes into the base of the thread used to screw the lid down

Neutral goes into the bar connecting the other side of the element

14 Put the lid on the new immersion heater and tighten up the bolt to keep it there.

15 You're now ready to put back all the fuses and turn the electricity on.

TIP *A lot of people get confused with how the hot-water cylinder controls the water temperature, so let's try to clarify things.*

If you use your boiler to heat the water then the cylinder thermostat will turn the heat off as soon as the water reaches the right temperature, regardless of how long you've set the hot water to be on for on your CH programmer.

If you use the immersion heater to heat up the water it will turn itself off as soon as its own internal thermostat has detected that the water's at the right temperature, regardless of how long you've set the immersion to come on for. As such there's no need to run up and down the stairs turning the immersion on and off all the time. Just leave it on and it'll switch itself on and off of its own accord.

If you're using low rate electricity, such as Economy 7, you just need to ensure that the immersion comes on and turns off during the low rate period. However, within this set time period the immersion itself will decide how long it needs to get the water up to the right temperature.

Boilers

First off, you shouldn't mess about with your gas or oil boiler. Play with water and get it wrong and you'll get very wet. Play with oil and get it wrong and you could be on the receiving end of an eye-watering clean-up bill. Play with gas and get it wrong and you could find yourself spread across four counties before you can say 'boom'.

At the moment it's not illegal to fit an oil boiler yourself, although you'll have to inform the building inspector and get everything checked and approved. However, it *is* illegal to fit a gas boiler if you're not a member of the Gas Safe register.

Modern room-sealed boilers are very, very safe. However, if you

don't understand all the safety features it's very easy to turn a perfectly safe boiler into a deathtrap. That said, it's also important that you understand exactly what's powering your home, and in Chapter 2 we discussed the type of CH system you might have. Here we'll briefly talk about the types of boiler you might have:

OPEN-FLUED BOILERS

Many old boilers are open-flued and floor-mounted, for no other reason than that they weigh a ton and few walls will support them. In order to burn their fuel such boilers take in some or all of the air they need from the room in which they're situated. Once burnt, the fumes just drift naturally up and away via a flue that's open to the room in which it's situated. From this description you can already see that these boilers can be very dangerous if things go wrong.

Over the years they changed the designs to make them less dangerous. They fitted fans to the flue to force the fumes out of the house and they fitted all sorts of detectors to the burner so that the boiler closes down as soon as there's any danger of poisonous gases being released.

Of course, these are only lifesavers whilst they work, so it's important to keep an eye on your boiler.

Ventilation

Because open-flue boilers take some or all of their air from the room you'll usually need additional ventilation. Sadly this also causes drafts, and it's not unknown for people to decide they'd

rather be warm and dead than alive and bothered by a slight draught; so the vents get covered over. Aside from the stupidity of deliberately blocking a vent they can also get blocked accidentally, by people putting furniture up against them.

The result of poor ventilation is poor combustion and the release of carbon monoxide, a tasteless, odourless, invisible gas that's thoroughly lethal to anyone who gets in its way.

So bearing all this in mind, always install a carbon monoxide detector.

Carbon monoxide detectors

Here are some general rules for carbon monoxide detectors:

- Fit a detector in every bedroom, where it can wake you if anything goes amiss.
- Fit a detector on every floor of your home.
- Fit a detector in every room where an open-flued or solid fuel burning appliance is located.
- Don't install the detectors directly above or beside boilers, fires or cookers, as these may emit a small amount of carbon monoxide upon start-up.
- Place them about 150mm from the top of the ceiling – you can go higher, but don't go lower, and keep them higher than any openable windows and doors.
- Keep them between 1.8m and 4.5m away from the potential source of CO, *ie* away from the boiler, fire, cooker etc, otherwise you get lots of false alarms and everyone starts to ignore the warning.
- Don't place in or very near to very humid areas such as bathrooms, as this often causes the detector to malfunction.
- If you have to clean them only use soapy water, as cleaning products can have an adverse effect on them.

Carbon monoxide: the warning signs

In terms of your boiler, the warning signs are:

- The flame in the boiler should be blue and well defined, with an inner and outer core. If the flame is flickering and is orange or yellow then it's not burning properly and is dangerous.
- When combustion starts to go wrong the flame will start to release soot, which will mark everything around the boiler. Look out for this telltale sign.
- In terms of yourself the warning signs are headaches, tiredness, dizziness and nausea. If you experience any of these around a boiler immediately turn it off and get it checked by a Gas Safe registered engineer.

ROOM-SEALED BOILERS

Because of the inherent dangers of open-flued boilers manufacturers dropped the idea altogether and started producing 'room-sealed' boilers. As the name suggests, the burner itself is separated from the house by an airtight seal and the boiler blows all the fumes and takes in all its air from outdoors.

The dangers here occur when the seal starts to deteriorate or if the flue terminates in an unsuitable position – it's a fat lot of good if the flue terminates outside but is just under a bedroom window where the fumes can get back into the house again.

These boilers come with a variety of safety devices designed to close the boiler down if anything goes amiss, but they still need an annual inspection to make sure all is well and that the boiler is working at maximum efficiency.

All modern boilers are room-sealed but there are variations within this classification:

Condensing boilers

These were designed with the words 'maximum efficiency' in mind. With older boilers the flue gases were often horrendously hot, which was not only a danger in itself but was also ludicrously wasteful – you spent all your hard-earned cash paying for gas and almost a third of the energy from it drifted up the flue. To rectify this, manufacturers started putting more than one heat exchanger into the boiler, so as to grab as much heat as possible from the gas being burnt.

You'd have thought that everyone would have been happy with this but sadly they weren't:

- Because the heat exchangers were much more efficient they tended not to last as long as the old fashioned cast iron exchangers and so the boilers had a shorter life expectancy.
- To fine tune everything and maintain optimum efficiency they used a Printed circuit board (PCB). Sadly, they tended to fail the moment anything went wrong with this board
- Finally, because they extracted so much heat back from the flue gases, much of the gas condensed and ran back down into the boiler as slightly acidic water, which was then piped away to a suitable drain.
- This worked a treat until we had a series of very cold winters and this pipe started to freeze up and close the boiler down.
- This was met with a great moaning and a wailing and a gnashing of teeth throughout the land and the tabloid press; some of whom chose to interpret this as a great socialist conspiracy, others as one meddle too many from the Eurocrates across the water.

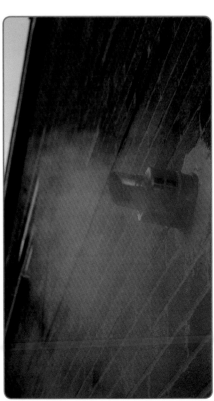

In fact this was all nonsense. What had actually happened was that the condensate pipe hadn't been fitted properly and, because we generally have mild winters, no one had noticed.

But all this is academic – if you buy a new boiler it will be a condensing boiler, and if you have any sense you'll be glad of this and the effect it will have on your fuel bills. Just make sure the condensate pipe is protected from any risk of freezing.

Combi boilers

The only reason for mentioning these here is that a lot of people get very confused about condensing boilers and combis. A condensing boiler is just a very efficient boiler. A combi boiler is one that generates hot water as well as central heating. 'Ah yes,' I hear you say, 'but don't all boilers generate heating and hot water?'

Well, yes ... and no. 'Heat-only' boilers only pour their heat energy into the central-heating system. External controls then divert this heat either to the radiators or to a hot-water cylinder. The boiler doesn't know where the heat is going and, frankly, doesn't care.

A combi, on the other hand, has two separate heat exchanges. One delivers heat to the central heating, the other delivers heat to the hot water. This has a number of advantages, in that the hot water is at mains pressure and, because it's instantaneous, you'll never run out of hot water whilst the boiler is working properly. The downside is that it can't generate very much hot water, and if your boiler breaks down you'll lose both heating and hot water.

System boilers

These are boilers that supply heat to a central-heating system that's under pressure. The advantages of this approach are that you don't need a feed and expansion tank in your loft; they're cheaper to install as a result; and the heat generated tends to get to the radiators quicker.

The downside is that you need to repressurise the system every time you bleed a radiator, which is done via a 'filling loop' usually located in or near the boiler itself – see Chapter 2 for details on finding and using the filling loop.

BOILER SUMMARY

Having read all this, it should be clear that you can own a 'room-sealed, condensing, combi boiler' or, perhaps, a 'room-sealed, heat-only, condensing, system boiler'. Whatever boiler you have needs to be serviced annually, not only to ensure that any warranties remain valid but to make sure that everything is running safely and that the boiler is working at optimum efficiency.

13 GOING GREEN

Insulation   203

Energy neutral homes   203

Underfloor heating   203

Solar thermal energy   209

Heat pumps   212

Biomass boilers   215

Rainwater harvesting   216

I was going to call this chapter 'alternative fuels' but everyone is getting in on the word 'green' these days so I thought I would as well, plus we'll be discussing technology that isn't an alternative to fuel.

There's a lot of nonsense spoken in the name of 'green', and I'm rather hoping not to add to it here, so let's just concentrate on what's important to us as homeowners.

First off, fossil fuel will, to all intents and purposes, run out. Some people say this will happen in 10 years time, some say 50 and some insist it will always be there. To be honest I don't really care, the fact remains that it will become rarer and rarer, will get more and more expensive to extract, and there will come a day when you might as well warm your home with bundles of £20 notes rather than use gas or oil.

So whether you believe in global warming or Peter Pan doesn't matter; gas and oil will become unaffordable to the average homeowner over the coming years, and there's not a lot we can do about that.

The other reason people have started to look at alternative fuels is because they care about the environment and want to do their little bit to mitigate some of humanity's actions. Like most good intentions this is a path fraught with difficulties, but it's an important area and an aspect of the technologies that we'll be looking at.

Insulation

The single most important aspect of heating your home is insulation. If your home is leaking heat to the great outdoors it doesn't matter what you're using as a heat source, you're still being incredibly wasteful. So before you ponder anything else check the following:

Do you have cavity wall insulation and is it sufficient? There's wild talk about the-powers-that-be taking thermal images of homes to see how much heat leakage they're showing, so I'd imagine that it will be possible to obtain a free report or at least hire this service in the very near future.

Check out the insulation levels in your loft and, if in doubt, put more in – they recommend at least 270mm of insulation in your loft these days.

If you haven't got double glazing get some. For listed buildings this can be an issue, but it's still possible to improve the insulation around windows. They do triple glazing these days and you can now buy glass with superb thermal properties. Yes, it costs more but it can also save you a lot of money over the lifetime of the window.

Check your floor insulation. This is an area often overlooked

when the house was built and by everyone else subsequently. As a result it's often the most poorly insulated area of the home and can therefore be easily improved.

 WARNING

Don't forget about radon gas. This is a radioactive gas that seeps up from the ground beneath your home. It's usually only a problem if you live in an area dominated by granite, but that's not always as obvious as you'd imagine. It's not good stuff, in fact there's talk about it being the largest contributor to lung cancer after smoking and, sadly, the more you insulate your home and cut out drafts etc the greater the danger of radon gas building up inside it. With this in mind, check to see if it's an issue in your area. If it is you might want to consider fitting a sump below your house to collect and dispose of this gas before it becomes a problem. In the meantime it makes sense to buy a radon gas monitor.

Energy neutral homes

These are also called zero-energy homes or zero net energy buildings (ZNE). We're not going to get pedantic about all this, so let's just say that they're buildings that leak next to no heat; what's more, they let very little in either. The result is a home that doesn't need a boiler because the heat from your own body, the TV, the cooker and every other electrical appliance in the home provides more than enough heat to keep the place warm, and in the summer you don't need any air-conditioning because the insulation also stops heat getting into the home.

All in all this is the perfect solution to fuel bills and is something anyone with an ounce of sanity should be aiming for. So what's the catch?

Well, to insulate an existing home to this level is an extremely expensive business, so much so that you might find it cheaper just selling up and buying a new pre-insulated home – although that's not to say you shouldn't try. All insulation saves money, you just need to get the balance right between money saved and money spent.

These new homes can be very expensive because they need to be built to a more precise standard, and the insulation materials themselves aren't cheap.

From an environmental standpoint the benefits are muddied somewhat by the insulating materials used, which are often petroleum based and as environmentally friendly as Jeremy Clarkson. However, there are alternatives now so you might want to check the building methods being used to decide if it's right for you or not.

The odds of us all being able to move into energy neutral homes tomorrow is very remote, so we're going to have to supply some sort of heat to our home in the meantime. So let's first look at how that heat might be used.

Underfloor heating

Most of the alternative methods of heating work best if your home uses underfloor heating. In traditional central-heating systems you attach a small metal plate to the wall (a radiator) and get it really hot in the hope that the heat will spread and reach every corner of the room. In underfloor heating you use the entire floor as a radiator and, because it's so much larger, you don't need to get it really hot. Also, because it's under your feet to start off with the heat generated rises evenly throughout the room, so you don't get cold draughts.

The important part of this description is the fact that underfloor heating doesn't get really hot; in fact the water within an underfloor system is generally around 35-40°C, as opposed to the typical 70–80°C, or thereabouts of a more traditional central-heating system. This is important for two reasons; firstly it makes it a better method for heating systems that employ a heat-pump (see later in this chapter for more details), and secondly the more you heat something, the more energy you tend to waste, so only having to heat the water to 40°C can already be a cost saving.

In reality the reason many people opt for underfloor heating is twofold; it's the ultimate in minimalist heating in that you can't see it at all, and it feels great to step out of a shower on to a warm tiled floor.

So what are the downsides to underfloor heating? Well, it can be more expensive to fit than good old-fashioned radiators, it's far more expensive and disruptive to repair if damaged, and it's really designed to work on tiled floors. Yes, it will work under wooden floors and some laminated flooring, but if you love thick shag-pile carpets then underfloor heating probably isn't for you.

THE OPTIONS

There's no right or wrong approach to underfloor, rather it's determined by what sort of flooring you already have, just how much of the home you want to convert to underfloor, and what your budget is.

Electric underfloor heating mats

This is the simplest way of introducing underfloor heating to your home and is probably the best approach if you're looking to heat a single small room – the bathroom, for example.

All you have to do is lay a mat of cables on the floor, introduce a thermostat to this web of cables, spread tile adhesive

over them and lay your tiles. You'll need an electrician to test it out and fit the fused spur but it really is very straightforward. That said, be careful not to lay the underfloor heating underneath things like shower trays, baths and toilets. And whatever you do, don't go drilling through the tiles afterwards.

TIP *Don't cut the underfloor cable to shorten it. This sounds obvious but it still happens. If you have excess cable left over you can often cut it away at the end that fits into the electrical socket, but even here there are limits. These systems work on the principle that electrical resistance generates heat – cut too much of the cable off and you lower the resistance, possibly to a point at which it no longer generates enough heat to warm the room.*

With this in mind it's very important to buy matting just large enough for your room – a 6m² mat for your 2m² bathroom is unlikely to work effectively.

You only want to heat up those tiles that can be walked on, and in a small bathroom this might not leave enough area to keep the room warm, so even with underfloor heating you might still need to add a small radiator.

In terms of energy efficiency they often talk about electric underfloor systems as the equivalent of leaving a light on in the bathroom all the time, although this demand will obviously increase with the size of floor area being heated.

Wet underfloor heating systems

If you're doing a large area, or want to convert your entire home, then this is the best approach to take. It's really just a matter of extending your existing central-heating system and there are really only two issues: how are you going to lay the pipework, and how are you going to control the temperature?

The first point is something we'll talk about shortly. The second point affects all 'wet' UFCH (underfloor central heating) systems. We need to keep the UFCH at a nice, steady, benign temperature – the last thing we want is to be dancing across the floor in bare feet yelling 'Ouch' because the floor is too damned hot.

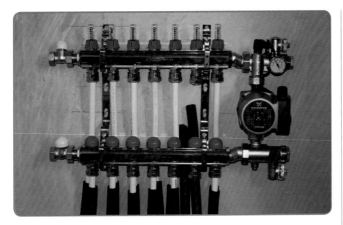

To achieve this steady, mild temperature we use a blending valve. In this arrangement our boiler continues to create water at about 80°C but this is kept away from the UFCH itself. Instead, the blending valve monitors the temperature of the water in the UFCH system and adds a small amount of the hot CH water to it every now and then to keep the UFCH system at about 35–40°C.

In small UFCH systems we can do this with nothing more complicated than a thermostatic blending valve. In larger systems we need what's called a manifold, which is a combination of a pump to circulate the water, a blending valve to keep the water at the right temperature, and a series of valves that allows us to turn off the underfloor heating in certain rooms and manage the rates of flow across the entire system. Manifolds can look horrendously complicated, but fortunately aren't.

In even larger systems it reduces the efficiency of the system for the boiler to turn on and off every five minutes in order to top up the UFCH. In this instance it makes more sense to add a 'thermal store' to the system – often called a 'buffer tank'. This is just a large body of water kept at the right temperature for the UFCH system and sits between the boiler and the manifolds. In this arrangement the boiler heats the thermal store and it's this store that provides heat for the manifolds. Because the thermal store loses heat very slowly it can keep topping up the manifolds for an age without the boiler running, which in turn saves energy. It's also very handy for systems that use an air-source heat pump, in that you can run the heat pump during the day when it's most cost-efficient, to heat up the buffer tank, and then use the hot water in this tank to supply the UFCH during the night, when the heat pump is running less efficiently.

UFCH FLOORING OPTIONS

Much of the sales pitch with regards to UFCH goes on about it freeing up the walls. This is true, but they neglect to mention that you now have to be pretty careful about your new radiator – the floor.

Tiled flooring

Underfloor heating works best if the final surface is hard enough to protect the pipework underneath and made of a material that's going to store and conduct the heat generated. Tiles achieve both of these objectives and are far and away the best surface to use with UFCH. But many people don't like tiles on their floors, so what else can you use?

■ Laminated flooring – this often works very well but you need to check that the laminate is appropriate for underfloor heating, as some isn't.

■ Solid wood flooring – this really depends on the thickness and type of wood. It's not really an ideal flooring as wood is actually a very good insulator, not the best property in what is, to all intents and purposes, a radiator.

■ Carpeting and rugs – not a good idea either. Putting underfloor heating down and then putting a carpet over the top is rather like fitting a radiator with a woolly jumper. You might as well forget about UFCH if you want wall-to-wall carpets, and if you opt for UFCH it's best to forego large, thick, rugs.

TYPES OF WET UFCH SYSTEM

Screed floor

UFCH works best if you set the pipework into something that will act as a heat store. When you do this the material around and above the UFCH absorbs the heat and then releases it in a slow, steady fashion. With that in mind, the ideal UFCH system is laid into screed and finished off with a tiled floor. Sadly, this approach is rarely possible in an existing home for the following reasons:

■ The floor of an existing house is usually not insulated enough to allow you to just lay the UFCH pipework directly on to it. So you need to add about 50mm of insulating material first.

■ The screed needs to be a minimum of 75mm thick to ensure it doesn't crack and break up as the UFCH system heats and cools. This minimum depth can vary depending on the composition of the screed and the pipework being used, so double-check with the manufacturer first.

So, to lay a screed floor in an existing home you either need to raise the floor level at least 150mm and learn to duck a lot, or dig out the existing floor and re-lay the whole lot. Either way is far from ideal and rarely cost-effective.

However, if you're building a new home or adding a large extension this is the perfect way to lay UFCH. Just make sure the architect and builder are aware that you want UFCH from the outset.

Low-level UFCH

Most manufacturers now offer some form of low-level underfloor heating product designed specifically for existing homes. The details vary but the basic concept is that you use a high density insulating material and lay

the pipework into it – sometimes this means the insulation comes with grooves built in, sometimes it means using an egg-carton-like matting that you lay the pipework into.

Most of these systems still require you to lay some form of insulation under them first, although this could be something as simple as a highly reflective surface to direct the heat from the UFCH upwards into the room.

THE DESIGN

Whilst laying underfloor pipework and connecting it to the manifold is well within the remit of the homeowner, the actual

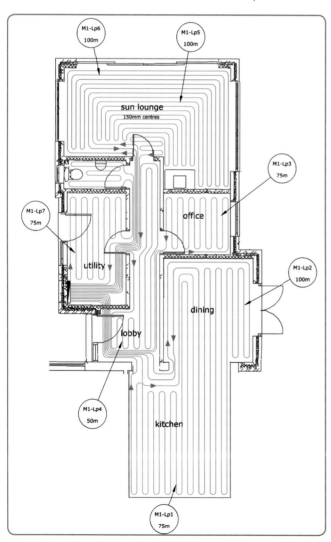

design and installation is best left to experts. Fortunately, most UFCH companies offer a design service with their product and it makes sense to take them up on this. They'll want to see a scale drawing of your home, the location of the boiler and/or the proposed manifold. They'll also want to know what your home is constructed out of so they can work out the heat loss – this usually means just telling them that you have cavity wall insulation and double glazing already fitted.

Once they have all that information they'll design the layout required and give you a list of the materials you'll need.

Be aware that most companies don't send you the actual layout until you've made a purchase. All you'll get initially is a materials list and a price.

In addition to this design service you might want to chat to a plumber to see what's the best way to provide the heat to the UFCH; directly to the manifolds, or indirectly via a thermal store.

LAYING THE PIPEWORK

TIP *Many UFCH systems now come with instructional DVDs as part of the kit. It might not be the most exciting DVD you'll ever watch but it should leave you pretty confident about how to go about installing your system. DVDs aside, each UFCH system is slightly different from the next so only treat the steps described here as general guidance – ie read the installation guide before you fit the system, rather than afterwards when it's all gone pear-shaped.*

The actual nuts and bolts of laying the pipework depend on what system you've opted for. However, it's relatively straightforward and the following points generally apply to all:

- Check that there's sufficient floor insulation. If this is a new build or an extension the builders will normally do this for you.
- The diagram supplied by the UFCH manufacturers will give you set pipe distances and show exactly what route each pipe should follow. It's important that you follow these instructions exactly. On low-level systems you're usually just pushing the pipework into the channels already cut into the flooring, so there's very little that can go wrong.

- On screed floors you usually clip the UFCH pipework to the top of the insulation. You can buy a tool for doing this, which will save you a lot of time and backache but costs a stupid amount of money for what it is.
- You want to avoid having any joints under the floor. As such the design should tell you what length of pipework you need for each loop (or circuit). The plastic pipe comes in rolls with the length marked on it, so check you're using the right sized roll for the loop you're doing before you start.
- Don't ever cross pipes over the top of each other.
- Most rooms will have a separate loop of pipework for just that room. However, with larger rooms there'll often be a series of separate loops, whilst for smaller rooms the loop will usually include the room and, say, the hallway.
- When you get back to the manifold make sure you label each pipe run and each connector – labels are usually supplied as part of the UFCH kit and you just jot down the name of the room, or rooms, serviced by the pipe. If more than one circuit covers a single room then either label them as 'Living room 1', 'Living room 2', or ideally be more descriptive and say 'Living room middle', 'Living room left side'.
- Most systems come with a number of wall-mounted room thermostats and a single floor stat. Make sure you put the floor stat in position before any floor surface goes down. Also, if given a choice go for the wireless room stats as they're much easier to fit and generate far less disruption to the rest of the house – although they're usually more expensive.
- When you come to the manifold make sure you don't bend the pipework so much that it kinks. To be honest the run up into the manifold usually looks like a right pig's ear as an array of plastic pipework emerges from the floor. Rather than compromise the UFCH with tight bends just put a decorative

cover over the manifold when you've finished everything. Some systems come with pipe bend-formers that allow you to click the pipe to the perfect bend – darned handy things that can make the run up into the manifold look almost good.

- All that remains now is to flush and pressure test everything. This is probably the most important step of all, so take the time to get it right and if in doubt ask a professional to do it for you.
- The system you've opted for should have clear instructions on how to perform this next step so make sure you follow them exactly. Bearing in mind that each system will be slightly different we'll just summarise the steps and principles involved.

1 For this procedure to work you need to ensure that water can only flow if it passes through your UFCH pipework, rather than taking a short cut via the pump. Every manufacturer seems to have a different way of achieving this; some supply a little Allen key that allows you to isolate the pump, some have a one-way valve built into the pump, and others insist that you perform this operation before fitting the pump. Check the manufacturer's installation instructions to determine their preferred method.

2 Firstly you need to isolate the manifold from the rest of the CH by closing the isolation valves on the main flow and return into the manifold (1 and 2).

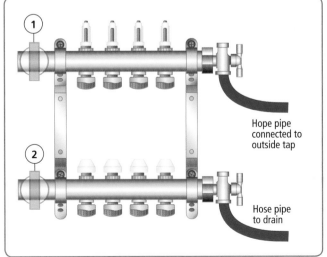

Hope pipe connected to outside tap

Hose pipe to drain

3 Next you connect the flow side of the manifold (usually the top) to the mains cold water and the return (usually the bottom) to a drain. If you have a single circuit you'll have no manifold. In this case just connect up the mains water to one end of the pipe and the drain to the other.

4 Each port on the manifold consists of two valves, one on the top and one on the bottom. A single pipe leaves from the top valve, goes around the UFCH circuit and returns to the bottom

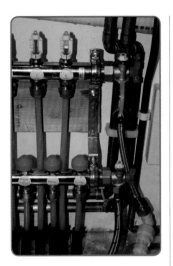

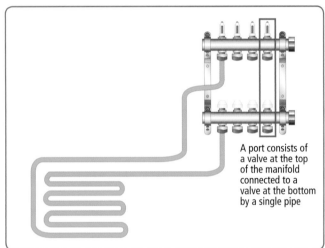

A port consists of a valve at the top of the manifold connected to a valve at the bottom by a single pipe

valve. You want to clean each circuit one at a time so you need to close all the ports except one. This one port will have a single valve open at the top and a single valve open at the bottom of the manifold. This is usually achieved by turning the cap on the bottom rail clockwise to close the valve and taking the cap off completely to open the valve – you'll notice that these look suspiciously like thermostatic radiator valves, which is no coincidence.

5 Run water through the first loop until the flow coming out of the drain hose is clean and flowing smoothly, with no sign of air in it. Once you're happy with this first loop, open the valves on the next port and then close those on the one you've just flushed.

6 Once you've cleaned out the circuits open them all up and connect a pressure tester to the manifold – usually at the filling point. Most hire shops will have these.

7 Raise the pressure in the entire UFCH pipework to 1.5 times the working pressure – usually 3–5bar.

8 It's best to be pessimistic about such things, so leave the system under test for an hour and then check the gauge. Because the plastic pipework will expand the pressure will drop a bit during this first hour. Bring it back up to the test pressure and then leave for another hour. There should now be no discernible drop in pressure. If there is, go look for the leak – on 99% of occasions any leak will be in the manifold connections and not the pipework, so start your search here.

9 After the test you can remove the pressure tester but you need to leave the pipework under normal operating pressure whilst the floor covering goes down. There are two reasons for this. Firstly, it stops the pipework collapsing under the weight of the flooring. Secondly, if the pipework is damaged during the screeding or floor-laying process everyone gets wet, so we get to know about the problem before it's too late to do anything about it.

10 Connect the room thermostats to the valves on the bottom rail. This is usually via a wireless control centre and is often best left to an electrician.

11 Set the flow rates for each circuit. This is like balancing the radiators of a standard central-heating system; you're setting

the level of resistance to flow to ensure that all the circuits in the manifold get enough hot water through them. For larger systems the design documentation you receive will include the appropriate flow rate for each circuit. If nothing is mentioned a flow rate of around 2 is usually deemed sufficient. Each system has a different way of setting these flow rates. On some you turn a nut on the bottom valves, on others you screw the top valves up and down.

12 For screed floors you need to leave the UFCH turned off for at least a few weeks so that everything can dry and settle without heat. Once all is dry and set, slowly bring the UFCH up to temperature over a period of days to stop the screed cracking.

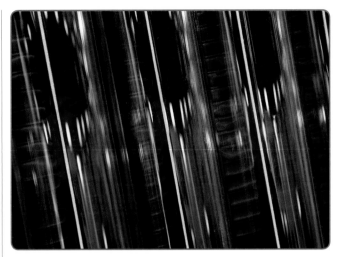

TIP

It often makes more sense if you think about the ports on a UFCH manifold as being the valves on a standard radiator. In this case the radiator is the single loop of pipework running under the floor. The bottom valve on the manifold is the thermostatic valve on your radiator, which will turn the radiator on and off according to how hot the room is, and the top valve is the lock-shield where you set the flow rate and then leave well alone. If you have a four-port manifold you effectively have a four-radiator CH system.

Our UFCH 'radiator'

Solar thermal energy

When we think about fuel alternatives solar is the one most likely to spring to mind. This is slightly ironic considering that most people regard the sun as being the natural phenomenon least seen in the UK. However, you don't need tropical beaches and sultry breezes to use sunlight as a source of energy, which is just as well.

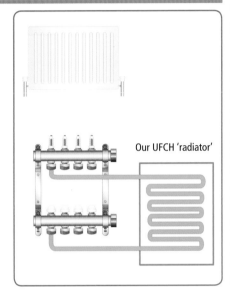

There are two mains forms of solar energy: 'photovoltaic solar panels', which produce electricity; and 'solar thermal panels', which heat up water. Photovoltaic panels lie firmly in the realm of the electrician, so we won't be discussing them here.

Before we go any further let's get down to the important stuff. How much money can you expect to save by investing in solar thermal, and is it really worth the bother?

Well, it will generally reduce your hot water heating bill by about 50%, which actually isn't an enormous amount of money. However, with gas prices going up in leaps and bounds the amount you'll save will increase proportionally and in October 2012 the RHI (Renewable Heat Incentive) scheme kicks in which should mean that the government give you around £250 per year for the heat your panels have generated.

Sadly, such grants have all the permanence of a heavy smoking mayfly with a heart condition. So you only *might* get money back for an uncertain period of time. That said, you are reducing your carbon footprint, and some of these panels now come with 25-year warranties, by which time natural gas will be so expensive they'll be selling it in gold-plated bottles.

So all in all it's regarded as a sound investment today and it's only going to become sounder as the years pass.

OK, so you should save money. But what system do you go for?

Well, solar thermal panels are divided into two main types: 'flat-panel' and 'evacuated tube' collectors. There's a lot of often heated debate about these approaches but very little by way of definitive conclusions.

FLAT-PANEL COLLECTORS

In its simplest form the flat-panel collector is just a metal plate under which runs a series of channels filled with water. The sun warms the metal plate and the water running through the channels picks up this heat and conveys it to your hot-water

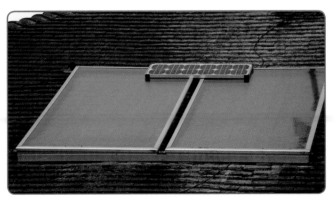

cylinder. If you live in the tropics this is fine, but if you like your water hot in the UK you may need to apply a bit more thought to your solar panel.

The paints, colours, shapes and materials used all affect the rate at which flat-panels absorb sunlight and convert it to hot water. As you might imagine, this is consequently an area that's in constant flux, and as such you're advised to refer to the website that accompanies this manual for up-to-date advice.

EVACUATED TUBE PANELS

In effect these work by taking a flat-plate collector and enclosing it in a vacuum – it's a bit more complicated than that, but in essence that's what's happening.

The great thing about a vacuum is that light can pass through it without difficulty but heat can't get past it at all. So once the sun's light has got through the vacuum and heated the plate up the heat generated can't get out other than via your hot-water cylinder.

With ordinary flat-plate panels the heat they gain is often lost back to the great outdoors before they have a chance to transmit it to your hot-water cylinder. This heat loss gets worse the colder the weather gets until you reach a stage where they don't work at all. The vacuum tube, on the other hand, still generates heat even when it's ice-skating weather outside.

There are two main ways of creating this vacuum. You can either take the vacuum flask approach, whereby the vacuum is created between two sheets of glass and these surround the heat plates, which are themselves in an area at normal atmospheric pressure (a glass–glass vacuum). Alternatively, you can put the heat plates and everything else inside the tube and make the entire thing a vacuum (glass–metal vacuum). The glass–metal approach is considered more effective, but it's harder to maintain the vacuum over time so they often don't have the same life expectancy.

An advantage of evacuated tube technology is that it often transfers heat from the tube to the rest of the system via a 'dry joint'. This means the tubes can be removed and replaced without draining the system. It also makes them more 'modular', allowing you to size your system more flexibly.

In fact the only real disadvantages of evacuated tubes are that there's more to go wrong and they tend to be more expensive.

COMPARING SOLAR PANELS

So which is best for you? Well, whilst we might not have the best weather in the world we also don't have the worst. British summers and winters are usually mild enough not to really need

a vacuum to protect the collectors from the cold. Yes, evacuated tubes are slightly better, and this performance difference increases as the weather deteriorates, but this is often offset by the fact that they're more expensive.

Over the coming years I'd expect evacuated tube technology to become better and better, whilst the price drops lower and lower. I'd therefore opt for evacuated tubes. Either way, you really need to focus on the balance between performance and price, *ie* if you can get evacuated tubes for a good price then go for it, but if you live on the south coast, have a south-facing roof and plenty of room on it then go for the cheapest option available.

Whatever type you buy you need to ensure that they've been tested to some sort of agreed standard. In the UK these are BS EN 12975 parts 1 and 2, which is a test of durability and performance. The performance test isn't a pass or fail, but it should come back with a percentage of efficiency, and if it did well the manufacturers will no doubt be eager to tell you all about it. Another advantage of BS EN 12975 is that most grants are only available if you fit panels tested to this standard, so you get a cost saving as well as an assurance.

Aside from the panels themselves there are a number of other 'system' issues:

Direct or indirect?

A direct system is where the water that comes out of your hot tap was recently running through your solar panel. These are far and away the simplest solar solution, but what happens if the water freezes? The answer is that is wrecks the solar panel, so such systems need a mechanism to turn the panels off when the outside temperature heads towards zero.

With this in mind, indirect systems tend to work better in the UK. These work by adding an antifreeze (food quality glycol) to the water in the panels. This is great for the panels but means that the water itself is no longer drinkable. To get around this you put a second coil inside your hot-water cylinder and run the water-glycol mix through this, which 'indirectly' heats the water in the cylinder.

Anything that works 'indirectly' is slightly less efficient, but being indirect means that your solar system is still delivering usable hot water in conditions where a direct system would have closed down. The only other downside of using an antifreeze is that it tends to break down over time and needs to be replaced, so you might want to find out the cost and frequency of this before you buy.

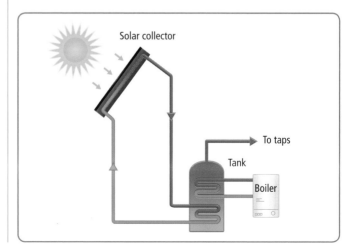

TIP *Many people assume that solar thermal panels rely on warm weather. This is not the case. What they rely on is access to the sun's rays. So if we have a warm, perpetually cloudy July and a freezing cold but bright January, most collectors will deliver more hot water in January.*

Passive or active?

A passive system uses gravity to do all the work but is consequently a lot more awkward to install. Active systems use a pump, so its various parts can be placed wherever's most convenient. As a result they're easier to install and often more efficient in terms of how well the overall system works ... but you're using electricity to run the pump, unless you opt for pumps powered by a small photovoltaic solar panel attached to the main solar thermal system. These are very effective but the pumps themselves are far from cheap and you might need more than one pump if your system is not south facing.

Pressurised (unvented) or open vented (drain back)?

By pressuring the solar circuit you increase the range of temperatures at which it can operate and have complete freedom to place components anywhere you like. They also tend to be quicker and quieter, and because they're sealed there's no danger of losing the water-glycol mix to evaporation. On the downside you have to fit various safety devices to cope with the high pressure.

As a general rule most solar panels on sale in the UK are indirect, active and pressurised systems.

SOLAR CYLINDERS

We've talked about how we get solar-heated hot water but we've yet to discuss what we do once we have it.

It's sad fact, but the likelihood of the sun being able to supply hot water to your home 365 days a year is remote. With this in mind you're going to have to use a combination of heat sources for your hot water – either your boiler and solar, or an electric immersion heater and solar.

The good news for people who currently use just an immersion heater is that solar will save them a fortune, and they might still be able to use their existing cylinder. But if you currently use your boiler to supply hot water then you're probably going to have to buy a new cylinder. Why? Because a standard cylinder only comes with a single coil inside it and you now need two – one for the solar system and one for the boiler to use.

That said, there's an alternative called the Willis Solasyphon. This is a neat little gadget that fits to the side of your existing cylinder and acts as the second solar coil. If your existing cylinder is the right size, well insulated and in good enough condition then this could save you a fair amount of money.

TIP *All solar hot-water systems need somewhere to store the hot water. If you have a combi boiler then your hot-water cylinder is probably lying in a skip somewhere, which means that solar thermal is not for you ... or is it?*

Well, if you bought a combi boiler so you could get rid of the

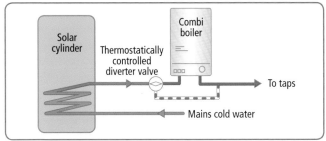

hot-water cylinder clogging up the airing cupboard, prepare to be disappointed, because it's going to have to make a comeback.

If you're determined to use solar and you have an old combi boiler your best bet is to buy a new heat-only boiler. If your combi boiler is brand new you could convert it back to heat-only, which isn't that expensive but does make buying the combi look like a waste of time.

A slightly more complicated alternative is to use the solar panel as a pre-heater. In this arrangement you fit a solar cylinder and connect its outlet to the cold-water inlet of the combi boiler. When the solar water isn't quite hot enough, eg in winter, the combi just heats it up that last little bit (which saves money), and in the summer – when the solar system can get the hot water up to temperature all by itself, – you simply use a little valve to divert this hot water past the combi direct to the taps. This is a neat little solution but does require that the boiler can accept pre-heated water, which isn't always the case.

The only other alternative is to keep your old cylinder, buy a new one and connect the two so that the solar cylinder acts as a pre-heater to the existing hot-water cylinder. The downside of this approach is that you end up living in the garden because the house is now filled with cylinders.

If you do opt to buy a new cylinder you need to make sure it's the right size for your home – see 'Sizing your solar thermal system' below. Other things you need to be aware of are:

- Insulation – all new cylinders must conform to the latest insulation standards. If your current cylinder is wrapped in lagging that's held on by string and a prayer, it's not going to work; throw it away and buy a new cylinder.
- Taller, slimmer cylinders work better than short, fat ones – this is because of the way hot water settles at the top of the cylinder (where we draw it off) whilst the cold water slouches at the bottom. Water does this whatever the shape of the cylinder, but in slimmer, taller cylinders the effect is more pronounced.
- Coil efficiency – most modern cylinders use longer coils with fins on them to maximise the rate at which they exchange their heat. With solar you might only have a limited time to exchange the heat so you need to ensure it happens quickly and efficiently.
- Solar coil position – the solar coil should be towards the bottom of the tank, below that of the boiler coil. This way the solar coil acts as a pre-heater for the boiler if there isn't enough sun to get all the water up to temperature.

⚠ WARNING

I've mentioned this before but let's mention it again ... There's a temptation to reduce the temperature of the water stored in the cylinder. After all, water at 60°C is too hot to bathe or shower in and the lower the temperature we store it at the more energy we save.

This is all true, but it overlooks bacteria. A pool of water at 40°C is single-celled heaven to some very nasty bacteria, eg Legionnaires' disease. These bacteria aren't killed off until the water temperature reaches 55–60°C, which is why that temperature is so important.

An additional issue with solar heated water is that the sun can't just be turned on and off, which means that you can't have exact control of the water temperature – it can get as high at 80°C in the summer. To resolve this you should fit either a single 'whole-house' blending valve to the cylinder outlet or fit blending valves to each hot tap outlet in your home to prevent scalding your family.

SIZING YOUR SOLAR THERMAL SYSTEM

The single most important way of ensuring you get value for money out of a solar thermal system is to size it properly. As a general rule of thumb you should assume each person in your house uses 40–60 litres of hot water each day. If you generally take a bath and have mixer showers aim at the higher figure, if you scorn the bath and always use an electric shower, then go for the lower figure. In addition you need to consider the amount of water that the panels are heating. As a general rule each square metre of panel requires about 80 litres of hot water storage capacity. Use both calcs to get a figure for the storage cylinder and then opt for the biggest value.

How large an area of solar panel you need varies according to the panels' position. Ideally you want to place them on a south-facing roof at an angle of 35°. Put them in this position and 1m^2

of panel is enough to generate about 45 litres of hot water per day, or roughly enough for one person.

As you move away from this ideal the efficiency drops, but fortunately there are umpteen calculators on the Internet to help you work out how much. Just type 'solar thermal sizing calculator' into your search engine. Try a few sites as everyone seems to use slightly different figures.

FITTING SOLAR PANELS

To be brutally honest this isn't worth doing yourself. Firstly you have to get the panels on the roof, which requires scaffolding. By the time you've hired the scaffolding, put it up and enlisted the help of friends it's probably not much more expensive to get a professional to do it for you. Whatever you do, don't try putting them up with ladders, as that's a recipe for disaster; at best you end up dropping the panel, at worse you end up dropping yourself.

On the bright side you usually don't need planning permission to fit solar panels. This changes if you have a listed building or live in a conservation area, but as a general rule if you've been allowed to put up an aerial or a satellite dish you'll have no problem putting up a solar panel.

The other reason for letting the professionals do the job is that grants are only available if the work is completed by a 'competent person', which currently means that they're an MCS (micro-generation certification scheme) accredited installer. This often affects the warranties as well; no accredited installer, no warranty.

It's also worth bearing in mind that the temperatures within a solar system can get very high. So high, in fact, that solder starts to melt and plastic pipework gives up the ghost altogether. Add to this the fact that most of the system is under high pressure and you really do have a recipe for disaster if things go pear-shaped.

Heat pumps

Heat pumps are nothing new. In fact the chances are that you have at least one heat pump in your home already, although they're usually referred to as 'the fridge', 'the freezer' and 'that bloody noisy air-conditioning unit'.

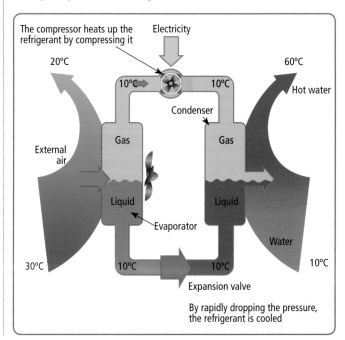

Heat pumps work by employing two fascinating facts of physics. The first one is that changing the pressure of something changes its temperature, and the second is that changing something from a liquid to a gas requires a surprisingly large amount of energy, which is released again when you turn the gas back into a liquid.

So if it's all simple old-hat technology, why all the fuss about air-source and ground-source heat pumps? Well, not only can a heat pump transfer heat from one location to another, it can also be used to magnify that heat. What's more it can achieve this very efficiently, so efficiently that for every kilowatt of energy you put into the pump you can transfer 4kW into your home! So whereas a traditional 5bar electric heater might be 100% efficient a ground-source heat pump could be 400%.

But this figure of 400% isn't as clear-cut as you might hope. The efficiency of a heat pump is determined by the temperature difference you're trying to achieve. If you take air that's freezing and try to supply heat to a radiator at 60°C you'll struggle to get much more than 100% efficiency. On the other hand if you're supplying water to an underfloor central-heating system at, say, 35°C and it's 15°C outside you may well get close to 300% efficiency.

Rather than talk about 200% or 300% efficiency the performance of a heat pump is measured in terms of COP – coefficient of performance. For example, a COP value of 4 is what most of us would refer to as 400%, COP 2.5 would be 250% efficient.

However, this COP value varies enormously depending upon the temperature difference you're trying to achieve, and it only measures the performance of the pump rather than the system as a whole. With this in mind you're better off looking at the SPF – seasonal performance factor. This shows the average performance of the entire heating unit, including additional pumps, fans, valves etc, over an entire heating season, *ie* during the year how much energy did the entire unit consume and how much did it generate? It is, of course, related to the COP value but gives the homeowner a much more realistic value to work with.

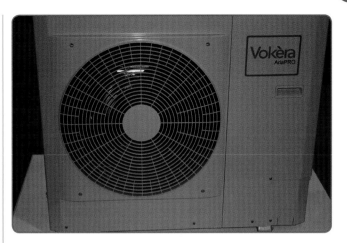

downside is that the temperature of the air fluctuates enormously, so that the pump is wonderfully efficient in the summer, when you don't need it, and not so efficient in the winter when you really do. That said, most good air-source heat pumps can maintain a COP value of 3–3.5 for temperatures between –3° and 10°C, which is as much as you'd expect from an average British winter, and some can maintain a COP of 2 for temperatures as low as –20°C.

TIP *The standard COP value given for an air-source unit assumes the air temperature is 21°C and that the CH system is running at 35°C. Whilst this output temperature is fine for underfloor central heating the air temperature value is wildly optimistic. Why would you want to turn the unit on when the temperature outdoors is 21°C? Fortunately, most manufacturers will supply graphs showing the COP or SPF of their units over a range of temperatures – between 2 and 7°C is more representative of an average UK winter.*

So what other things do you need to consider when you're looking at air-source heat pumps?

Sizing your air-source heat pump

Sizing is the most important factor. You need to know how much heat your home requires to keep it warm in winter and size the heat pump accordingly. Of course, the amount of heat your home requires depends upon the outside temperature; the colder the weather, the more heat you'll need. With this in mind you need to set a design temperature and work out the heat loss at this temperature.

In the UK we traditionally use –1°C to design the heat requirement for boiler driven central-heating. 'But,' I hear you say, 'what happens when we get a winter colder than that?' Well, the short answer is that you get cold or you use additional heating.

The problem is that you have to set a realistic temperature. It's pointless designing a system for temperatures of –15°C when you can live in the UK for decades and never experience a temperature even remotely close to this. If you did use –15°C you'd end up with a vastly oversized system which was highly inefficient, except for one day every decade.

Most companies selling air-source pumps will calculate your heat requirements as part and parcel of you buying the unit. However, you can get a vague figure yourself using this calculation:

TIP *The lower the output temperature of a heat pump the more efficient it is. This is why heat pumps work better with underfloor heating – UFCH operates at 35–40°C whilst radiators need 60–70°C. However, it's not always cost-effective to fit UFCH throughout your home, so a compromise is to oversize your radiators and run the entire system at a lower temperature. Dropping the temperature to 50°C and increasing the radiators by about 30% will generally work. If you opt for this route you need to mention it to the manufacturer, as it'll affect the efficiency and size of the heat pump unit.*

The other thing to look for is EN14511. This is the standard European test for heat pumps, and it's not a smart move to buy a pump that hasn't been through this test.

The two most common heat source pumps are air and ground.

AIR-SOURCE HEAT PUMPS

It's actually quite difficult to tell the difference between an air-source heat pump and an air-conditioning unit. This is because they're pretty much the same thing but working in reverse.

Using air as a heat source has the advantage of being relatively cheap, in terms of both installation and maintenance. The

Category
Which category best describes your home?
New build = 50
Older home with cavity wall insulation = 70
Old house with no insulation = 100

Location
Scotland
England (North and Midlands) and Wales
Southern England

Area of your house
Calculate the area of the ground floor of your home in square metres.

Radiators or UFCH
Now decide if you want to use oversized radiators or fit underfloor central heating.

To get an idea of the heat loss of your home take the 'Category' figure, multiply it by the 'Area of your house' and divide by 1,000.

For example, I have an older home with cavity wall insulation that's got an area of 120m², so the calculation is 70 x 120/1,000 = 8.4.

Now find this 8.4 in one of the following tables, depending on whether you're planning on using oversized radiators or underfloor central heating:

Location	Air-source unit size (using radiators)				
	6kW	8kW	9kW	12kW	13kW
Scotland	3.9	5.8	7.1	8.5	11.3
North/Midlands/Wales	4.1	6.0	7.4	8.9	11.6
Southern England	4.3	6.4	7.7	9.4	12.1

Location	Air-source unit size (using underfloor central heating)				
	6kW	8kW	9kW	12kW	13kW
Scotland	4.0	6.0	7.1	8.8	10.5
North/Midlands/Wales	4.2	6.3	7.4	9.2	11.0
Southern England	4.5	6.7	7.8	9.8	11.5

In our example we're looking at Newcastle in the North of England and we're using radiators; so we refer to the middle row of the upper table.

There isn't an 8.4, so we go for the next highest value for the North/Midlands/Wales, which is 8.9 and equates to an air-source unit of about 12kW.

NOTE *Many people think that a heat pump won't work if the air temperature drops below zero because there's no heat to be got from the air at this temperature. Yes, the efficiency of a heat pump will drop, but they'll still work down too much, much lower temperatures. In reality everything has 'heat' in it providing it's warmer than –273°C. Looked at this way 0°C is actually pretty hot!*

Remember that this still means there'll be very cold days when the heat pump can no longer supply enough heat. So what do you do on those days? Well, you could fit two smaller heat pumps and design things so that the second pump comes to life when the temperature drops below a certain level. This results in a pretty efficient system but isn't cheap! An alternative is to leave your old boiler in place and design the system so that this boiler automatically kicks in when the weather drops below a certain temperature.

Most units come with a standard electric heater already built into them. This is an expensive way of heating your home, but if the system has been designed properly you'll only use it once in a blue moon.

Locating an air-source heat pump

Where you place the unit is going to have an enormous bearing on how efficiently it works. First off you don't want it swamped by dead leaves and snowdrifts, so it's best located at least 400mm off the ground.

You should site the unit facing south and in full sunlight. This has less to do with the air temperature and more to do with the unit temperature. Humid air will deposit ice on the unit at temperatures between 0–5°C, but this is less likely to happen if the unit is being warmed by the sun.

TIP *A UK winter is ideal for depositing ice as the air is often both humid and cold. To remove this the air-source unit goes into 'defrost' mode and at this point its efficiency drops through the floor. Modern units use an array of devices to reduce this effect but location is still the most important factor.*

Since air is the source of all the heat we're going to generate the unit has to have access to plenty of it. So it should be outdoors in a reasonably open location. That said, you don't want strong winds whipping across the unit all the time so shelter from prevailing westerly winds is an advantage. What you should avoid is running the unit where there's little access to free air. In this scenario the heat pump will just start to act as a fridge, reducing the temperature of the stagnant air and becoming less and less efficient as it does so. That said, they actually design air source heap pumps for location in the loft! Although I assume the loft needs good ventilation and that this location is picked for convenience rather than performance.

GROUND-SOURCE HEAT PUMPS

Ground-source heat pumps take heat from the ground and pass it on to your home. The big advantage of ground-source is that, past a certain depth the ground tends to stay the same temperature

Courtesy of Dimplex Renewables

regardless of the weather. As a result a ground-source heat pump with a COP of 4 in the summer will probably still have a COP of 4, or thereabouts, in the winter.

NOTE

People often get confused between 'geothermal' and 'ground-source'. Geothermal is very popular in places like Iceland, where volcanic activity is commonplace and the ground is often very hot as a result. Ground-source is a much calmer system. It utilises the fact that the sun warms the ground throughout the summer and that the first few metres of soil are relatively snug as a result. The temperatures are nothing compared to those of a geothermal system but at least you don't have to contend with lava erupting all over your lawn.

To get the heat out of the soil you need to place some sort of heat exchanger in the ground. This can be something as simple as a series of plastic pipes thrown into a metre-deep trench and looped around and around or it could be plastic 'radiators' buried into your lawn or, if you don't have a lawn, it could be a borehole drilled directly downwards – though the cost of digging a borehole is prohibitive, and you might not even get permission.

Once the pipework's in place you generally fill it with brine or a mix of water and glycol ... and that's all there is to it! You pump this mix of water and glycol around the circuit you've just laid and the heat pump grabs the warmth out of it. It then magnifies this heat and passes it on to your central heating.

The only real downside of a ground-source heat pump is the initial cost of installing it. On the plus side the underground circuit should last at least 50 years and the heat pumps themselves will often last about 25 years, so they'll certainly pay for themselves eventually.

There are a number of factors that affect performance. For example, the damper the ground the better the system tends to work, because water is an even better heat store than earth, and you tend to get better heat transfer if the pipework's immersed in water. Most important of all is sizing the system correctly, which in turn depends on your heat need and the soil type. All of this should be taken into account by the installers when they do their initial survey, and before you ask, yes, it's best to leave all this to the professionals unless you happen to have a degree in geology, an A-level in maths and a part share in a JCB.

ARE HEAT PUMPS GREEN?

Sadly, the answer to this is often 'no'. This isn't the fault of the heat pumps, rather it's the fact that they're using mains electricity.

The problem with electricity taken off the national grid is that it's produced by clapped-out coal, oil and gas-fired power stations. Not only are they burning fossil fuels but they're burning them very inefficiently – by the time the electricity reaches your home the efficiency level has dropped to about 20%! As a result the carbon footprint of mains electricity is a mammoth size 17.

By being super-efficient all that a heat pump can do is get the carbon footprint of the electricity down to a size where it's comparable to that of natural gas. In terms of price, the situation is very similar; electricity is still more expensive than any other fuel, but because it uses it very sparingly a heat pump can get the running cost down to about that of a new natural gas powered boiler.

However, if you buy your electricity from a green supplier or generate your own via photovoltaic solar panels (PV), heat pumps suddenly take on a hue comparable to that of Kermit the frog on a diet of spinach.

You can't use the PV panels directly because they tend to work best in summer when the pump isn't working, and they can't supply the starting load required by the pump. However, it's possible to cover the costs of running your heat pump by selling your photovoltaic electricity back to the national grid, and you don't need an enormous amount of PV panels to do this.

Manage this and not only are your heating costs zero but you can hug a tree with a clear conscience.

Biomass boilers

Biomass is the generic term for boilers burning wood pellets, logs and anything else made out of tree or plant material. Again this is nothing new and the Scandinavians have been burning wood pellets from sustainable forests for an age.

In the bad old days of wood-burning furnaces you needed someone to add the logs on a regular basis and someone to clean away the ash afterwards. With modern biomass boilers there's very little ash left over because they burn so efficiently. So you only need to empty the ash tray every month or so, and some boilers will even do that bit for you. Most also come with automatic feeder systems so it's just a matter of filling a big hopper every now and then and leaving the boiler to do the rest.

So how are they green? Well, trees and plants absorb carbon dioxide during their lifetime and release it when they're burnt or allowed to rot. By burning them in a boiler all you're doing is releasing back to the atmosphere the carbon dioxide the tree absorbed a few years ago and would have released anyway if allowed to die of old age. So there's no net increase in carbon dioxide levels.

Of course there's some net increase, because you need to process the wood and then transport it to your home, but on the whole they have a very small carbon footprint.

The only real problem with biomass as a common approach to

Keeping the heat exchanger clean on a bio mass boiler.

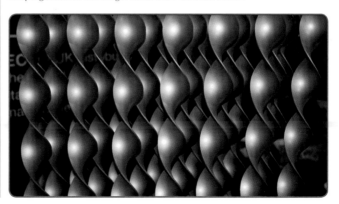

home heating is the question of sustainability. The Scandinavians have enormous forests and a relatively small population. We, on the other hand, have very few forests and an enormous population. Whilst biomass demand is relatively low it works a treat, but if we all decided to convert to it we'd run out of woodland by Tuesday.

NOTE *Wood-burning stoves and biomass boilers aren't the same thing. A wood stove is not a highly efficient way of heating your home – it is a fire in a box – although they can look quaint, if you like that kind of thing. Again they're considered green because they burn sustainable wood that only absorbed the carbon dioxide the fire emits a few years ago. But they're not as green as a biomass boiler because they don't burn the wood anywhere near as efficiently.*

Another issue to consider is storage. Biomass is, on average, about 14 times bulkier than oil, so that to heat your home for the same length of time you'd need 14 times the storage area. This means either setting aside a larger area for storage or arranging for more regular deliveries, which is expensive and degrades the green credentials of the biomass product.

If you're going to process wood or organic material yourself you need to pay a lot of attention to the storage process, as getting the moisture content right is very important if it's to burn correctly. This can mean storing the wood for a long period of time, which of course means even larger storage areas.

Size is also a problem with the boilers themselves. It's not so much the boiler as the hopper mechanism, which can be very bulky.

All in all this is a very good approach for a niche market but it's not a viable long-term option for all of us. That said, any landowner with access to sustainable woodland or arable by-products has to be mad if they haven't yet looked at biomass as an alternative fuel source. Yes, it's not cheap to convert, but there are grants available and a biomass boiler will almost certainly pay for itself over its working life, during which time you won't be adding carbon dioxide to the atmosphere and can hug a tree with a clear conscience – assuming you haven't already chucked it in the boiler.

Rainwater harvesting

The nation's water companies spend a great deal of time and trouble ensuring that the water that arrives in your home is wholesome and wholly devoid of bacteria and other nasties. To make sure we stay fit and well, and have a sparkling smile too boot, they often add fluoride too. And how do we repay them for this service? We flush it down the loo!

About 26% of all water used in the average UK home is used to flush the toilet, with a further 19% used to wash clothes, water the garden and clean the car. This is enough to break the heart of even the hardest water treatment plant, and not only is it daft but it's a waste of money. It's on this basis that rain and grey water harvesting has risen in popularity over recent years.

In rainwater harvesting you collect the rain off your roof rather than letting it run away into the sewer or drain away into the ground. In grey water harvesting you usually still collect rainwater but in addition you collect the waste water from your shower and bath. You can collect the waste water from your washbasin and kitchen sink too, but these are often have all sorts of dubious materials down them and as such require either far more complex treatment or a strict policy on what can or can't be poured down the sink.

So what can you use the water for? Generally speaking rainwater can be used to flush the toilet, water the garden and wash clothes, whilst grey water is usually only used to flush the toilet. It's possible to really push the boat out and render your collected water drinkable, although this involves numerous filters, including a reverse osmosis system, and probably ultraviolet treatment, to say nothing of regular checks and inspections to make sure nothing's gone awry. All in all it's not for the faint-hearted, and from an installation point of view is definitely best left to the experts.

TIP *Rainwater has already been softened by nature, as such you need less detergent in your washing machine and there'll be no scale build up in your toilet.*

THINGS TO CONSIDER AT THE VERY START

The best way to install a rainwater harvesting system is to sit down with a sheet of paper and plan a new house with a rainwater harvesting system integrated into it from the outset. That's not to say they can't be retro-fitted, just that it's not always feasible and not always cost-effective.

If you are looking into this for your home there are a few things you need to think about:

Do you have a water meter?

If the answer is no, you're not going to save any money by installing a rainwater harvesting system. However, you may well be reducing the pressure on local reservoirs, lakes and rivers. They also reckon that you reduce the risk of localised flooding, presumably by storing water that would normally run off your roof and join the flood.

If you're on a water meter you should be able to reduce your bills by about 50%, so a cursory glance at your water bill should tell you if the expense of a rainwater harvesting system can be justified.

How much water do you need to store?

As a rule of thumb assume that each person in your home will use about 20 litres of water a day to flush the toilet. So calculate your requirement either based on how many people live in your home or, probably a better approach, how many people would you generally expect to have living in your home based on the number of bedrooms you have.

How many continuous days without rain does your area generally have?

What we're trying to calculate here is how much water will we have to store in order not to run out of rain water, *ie* we need 20 litres a day and it often doesn't rain for 10 days at a time, so we'll need to store at least 200 litres to get us through these regular droughts.

If you live in North Wales you're probably more used to calculating this amount in minutes rather than days, but other areas of the country can get regular and prolonged dry spells. To find out this information go to the Met office website (www.metoffice.gov.uk) or ask the manufacturers of your rainwater harvesting system.

These factors enable you to work out your requirement as follows. Let's assume you have a three-bedroom house that would normally have about four people living in it. Let's also assume you live in a fairly dry part of the country, where periods of 25 consecutive days without rain would be the yearly average. We therefore need to calculate litres per person x number of people x average drought period in days, which in our example would be 20 x 4 x 25 = 2,000 litres.

How much rainwater can you actually collect?

There are two main factors here:

Location

Rainfall across the UK varies dramatically. As a general rule it's wettest in the west and driest in the east, but to size a tank properly you need to find out the average rainfall for your area. You can usually discover this information at the Met office website.

Collection area

The design of your home will generally dictate how much of the water landing on it can be collected. If you live in a terraced house it's going to be nigh on impossible to collect the water from both sides

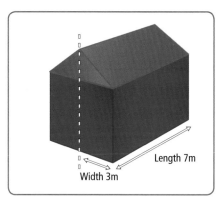

Length 7m
Width 3m

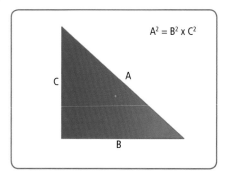

$A^2 = B^2 \times C^2$

C   A

B

of the roof, but even in a detached house it's probably going to be impracticable to divert all the rainwater downpipes to a single location. With that in mind, choose some possible tank locations and then examine what rainwater pipework can be diverted to this location and just how much roof area this involves.

To measure the area of roof you're going to use it's usually sufficient to measure the width and length of the building it sits on in metres. To work out the square metreage multiply the length by the width. For example, in our diagram we have a length of 7m and a width of 3m. This gives us 7 x 3 = 21m².

If you have a very steep angled roof, or are a stickler for accuracy, you might want to calculate the width of the roof using Pythagoras' theory to calculate the hypotenuse. You can measure B by measuring the width of the house and dividing by two. You can calculate C by counting the number of bricks from the base of the roof to the top of the pitch and then measuring the distance between the base of one brick and the base of the next – always remembering to calculate all your measurements in metres.

With these two factors you can calculate the amount of water available for collection thus:

Annual rainfall (m) x effective collection area (m²) x 0.8 (we usually only collect 80% of the water landing on our roofs)

So if you live in North Wales you might get 1.5m x 21m² x 0.8 = 25.2m³. There are a 1,000 litres in a cubic metre, so this converts to 2,520 litres, which is the amount of water available for use, whereas we earlier calculated that we needed 2,000 litres.

It's rare for both calculations to come out with the same value, so you should always take the lower value when selecting the size of the tank, 2,000 litres in our example. This means you should rarely if ever run out of rainwater but also means that the water's unlikely to stagnate and the tank will overflow occasionally, which is needed to remove debris.

Of course, if the actual amount you can collect is much lower than the amount you need you might want to have a rethink about rainwater harvesting altogether.

How will you get the water to your toilets, washing machines and taps?

The pipework containing your rainwater cannot under any circumstances be allowed to mix with the hot and cold water

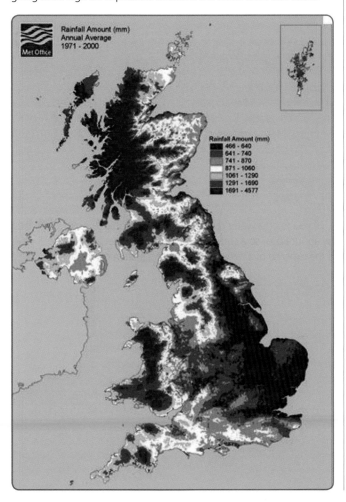

Rainfall Amount (mm)
Annual Average
1971 - 2000

Met Office

Rainfall Amount (mm)
466 - 640
641 - 740
741 - 870
871 - 1060
1061 - 1290
1291 - 1690
1691 - 4577

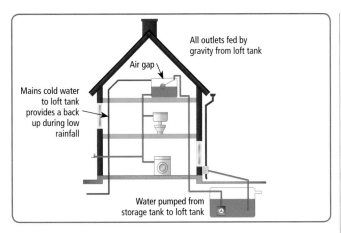

All outlets fed by gravity from loft tank

Air gap

Mains cold water to loft tank provides a back up during low rainfall

Water pumped from storage tank to loft tank

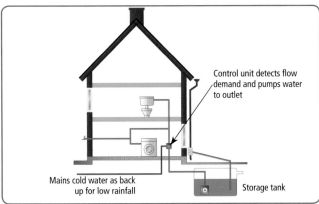

Control unit detects flow demand and pumps water to outlet

Mains cold water as back up for low rainfall

Storage tank

Other things you need to consider are:

■ It's best if the tanks are buried underground. This keeps them cool and dark and so reduces the chances of bacteria and algae building up.

■ Do you have a large enough garden to accommodate this tank? Is it going to be possible to get the tank *into* your back garden?

■ It's possible to collect water from driveways etc but such water is often contaminated by all sorts of unsavoury substances so you'll need additional filters and water treatments.

■ What's your roof made from? You can collect water from asbestos roofs but they can readily clog filters and there's a potential health hazard. Metal roofs will often stain the water and bitumen or felt roofs can cause colouration and odour problems.

■ How high is the water table in your garden? If the answer is high you'll need to either line the hole for the tank with concrete or fit some sort of restraining system to stop the tank floating away every time it rains. Alternatively you might want to look at collecting this ground water, which is an option available with some tanks.

SELECTING A RAINWATER HARVESTING SYSTEM

At its simplest a rainwater harvesting system is nothing more than a water butt. However, in reality there's far more to it than that. There's now a British standard for rainwater harvesting systems called BS 8515. This is a code of practice rather than a set of standards but all reputable companies will follow it and will advertise this fact on their products. In other words if you don't see BS 8515 don't buy.

■ You need a series of filters, ideally ones that require little by way of maintenance, to remove any particles and general debris from the water. By sizing the tank correctly you should be able to float a lot of the debris away during periods of high rainfall but the filters will still need to be maintained, on average one to three times a year depending upon the tree coverage in the area. Have a look at the tank design to see how simple a process this is going to be.

supplying the rest of the house. It must run in separate pipework, and this must be clearly marked as not being drinking quality water. The outlets must be similarly identified, though this is largely a waste of time as the only creatures likely to drink out of the toilet can't read. However it's always best to play safe, and your dog might surprise you with its grasp of English.

Where the toilets, washing machines and outside taps are will dictate just how much new pipework will need to be laid, and if it's actually going to be feasible at all.

Another consideration is how are you going to get the water into these pipes in the first place? You could opt for a harvesting system that moves the water out of the main storage tank and into a smaller loft tank, from where it runs by gravity to all the outlets. Or you could opt for a direct system, were the water's pumped directly from the main storage tank to the outlet. Again, the layout of your home might dictate which approach is going to work best, although as a general rule the pumped system is easier to install.

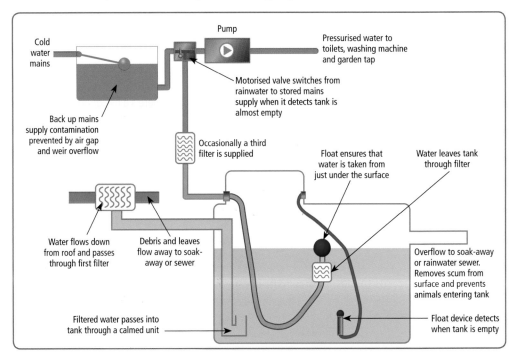

Cold water mains

Pump

Pressurised water to toilets, washing machine and garden tap

Back up mains supply contamination prevented by air gap and weir overflow

Motorised valve switches from rainwater to stored mains supply when it detects tank is almost empty

Occasionally a third filter is supplied

Float ensures that water is taken from just under the surface

Water leaves tank through filter

Water flows down from roof and passes through first filter

Debris and leaves flow away to soak-away or sewer

Overflow to soak-away or rainwater sewer. Removes scum from surface and prevents animals entering tank

Filtered water passes into tank through a calmed unit

Float device detects when tank is empty

- If it hasn't rained for a while the odds are that the first downpour will bring with it a lot of debris. The first filter should be designed to allow this debris to float away before it starts to divert water into the storage tank.
- If you just pour the water into the main tank it's going to stir up sediment, so you need a 'calmed' inlet that allows the water to flow into the tank in a smooth fashion.
- You don't want to suck up any sediment or surface scum so the inlet to the pump should float just under the surface. The pump itself can either be in the tank or in a separate unit. It's usually easier to detect pump problems and service a pump that isn't at the bottom of a large water-filled tank.
- To use the water for your toilet flush and washing machine you need to get the water from the butt and into the house at a usable pressure. To do this you can either pump the water into a loft tank, or pump the water directly to the outlet. Pumping into a loft tank allows you to use low tariff electricity but does add to the purchase and installation costs.
- You can't rely on rain and you can't only use the toilet during times of deluge. With this in mind you have to have a mechanism whereby the system can be topped up with mains-fed water whilst ensuring that the mains water can't be contaminated. To do this automatically the tank should detect when it's almost empty and switch to using mains water. You want this switch-over to be automatic but it's handy if you also get some sort of warning so that you know when you're on mains water.
- You need an overflow in case you get too much rain. This can be directed to a soakaway or reconnected to your main rainwater sewage pipes, but should include a rodent barrier.
- You'll need to be able to remove the sediment that will inevitably build up over time. Ideally this will be a clean, simple process that you don't have to do more than once a year. It's a good idea to compare the servicing instructions of different systems.
- Your guttering will need to be cleaned regularly to ensure the rainwater remains fairly clean and free from 'things'. The main danger of using a rainwater harvesting system is the potential for exposure to pathogenic micro organisms (germs) and faecal matter (bird poo). The best way to keep these under control is to keep the entire collection system clean.
- So, is rainwater harvesting green? Well not according to the Environment Agency. As they point out, the green thing to do is reduce the amount of water you use, rather than just collect it from a different source. The other problem is that you're using electricity to drive the pump, and as we've already mentioned, electricity isn't even vaguely green by the time it arrives in your home. You could design a rainwater harvesting system that doesn't need an electrical pump, but this usually requires that you design it when you design the house.

GREY WATER HARVESTING

You can get really serious with rainwater and grey water harvesting and, in effect, operate your own water treatment plant. However, that would be a book by itself so we'll just look at the simplest approach.

At its simplest the waste water from your bath and shower is fed into a storage tank – you could take waste from washbasins and sinks too, but you'd need to ensure that things like oil and food waste weren't washed down the drains.

Once in the storage tank the waste sediments out. The lighter

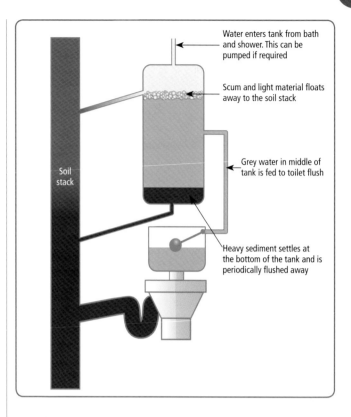

Water enters tank from bath and shower. This can be pumped if required

Scum and light material floats away to the soil stack

Grey water in middle of tank is fed to toilet flush

Heavy sediment settles at the bottom of the tank and is periodically flushed away

Soil stack

scum floats to the surface and runs away down a drain. The heavier sediment falls to the bottom of the tank and is periodically flushed away to the soil stack. In the middle of the tank you now have relatively clear water, wholly unsuitable for drinking but perfectly fine for flushing a toilet.

In reality things are a bit more complicated. For example, you can't store this water for more than 26 hours without risking a build-up of bacteria, so the system will have to automatically flush itself before this limit is reached.

You also need a back-up of mains cold water for times when grey water's unavailable. This would be fed from a tank, via an air gap and a weir overflow, and would usually be built into the system.

The upside of these systems is that there's no need for a large tank to be buried in your garden, so the overall installation cost can be lower. The downside is that you don't collect as much water and there's a risk of odours developing from the flush water itself.

Of course, there's no reason why you couldn't integrate grey water and rainwater harvesting, with the grey water supplying the water for the toilet and the rainwater being used for the garden and washing machine.

GRANTS

All grant schemes seem to work around a fine balance; on one side they have to offer an incentive for positive change, on the other hand they don't actually want to give any money away if they can possibly help it.

At the time of writing the main grant for green energy is the RHI (renewable heat incentive). How long it will be around and how much money will be available is anyone's guess. You can keep abreast of grant developments by looking at the website that accompanies this manual, or the umpteen websites that pop up when you enter 'green energy grants' into your search engine of choice.

Glossary

ABS – Acrylonitrile butadiene styrene. A plastic commonly used in plumbing for waste pipework. Sadly it doesn't cope with sunlight very well unless mixed with a UV stabiliser, so it's better to use MUPVC outdoors.

Airlock – A bubble of air that gets lodged in water pipes and effectively blocks them. This is most common in gravity-fed hot water systems where there isn't enough pressure to push the bubble out of the way.

Ballcock – Device invented by Thomas Crapper, composed of a valve and a float. As the water level drops, the float falls and the valve opens; as the water level rises, it lifts the float which closes the valve. Also called a 'float valve' or a 'ball-valve'.

Ball-valve – An alternative name for a ballcock. Also used to describe other spherical valves, such as isolation valves, where a ball bearing with a hole through it is used to control water flow.

Basin spanner – Also called a basin wrench. An essential item of plumbing equipment, as it allows you to undo nuts located in tight spaces where no other tool can get access.

Biomass boiler – A boiler that burns biological matter from recently living organisms. Items such as logs, wood pellets, hay, straw, maize etc are all classed as biomass.

Bottle trap – A trap is a U-bend that uses a trapped volume of water to stop smells emerging from waste pipework. In a bottle trap the U-bend is formed by having a small tube inside a larger one. They look neater than other traps but shouldn't be used on kitchen sinks, as they tend to block easier.

Box spanner – A metal cylinder shaped to encase a particular-sized nut. They're very useful when you can only access a nut from directly below.

BSP – In order to fit a nut on to a bolt you need them to be the right size and have the same type of thread. BSP is the British standard for pipe threads, so if you have a 15mm nut and a 15mm bolt they'll always fit if they adhere to the BSP standard.

Btu – British thermal unit. Used to describe the heat output of boilers and radiators, although now largely regarded as old-fashioned and has been superseded by the kilowatt (kW): 1kW equals 3,412btu.

Byelaw 30 kit – Kit used to insulate and protect loft storage tanks.

Captive-plug waste – Rather than hanging the plug on a chain or popping it up and down with a bar, the captive-plug system keeps the plug trapped in the plughole but allows you to flip it. When flipped on edge water runs out of the basin; when flipped flat the basin can be filled.

CH – Common abbreviation for central heating.

Check valve – A valve that only allows water to pass through it in one direction.

Clicker waste – A neat little plug that's fixed into the waste of the bath or basin. You press it down once, it clicks and the plug is closed. Press it down again, it clicks and the plug is open.

Coffin tank – Long, thin, low storage tank that allows a lot of water to be stored in a loft but will also fit through most loft hatches. The shape has more than a hint of the coffin about it, hence the name.

Collet – A tapered collar that's used to clamp something into place. The most common collet in plumbing is affixed to push-fit fittings – push the collet down and the pipe is released, let it spring back up and the pipe is held firmly in place.

Combi or **combi boiler** – Abbreviations for 'combination boiler', a boiler than generates hot water as well as powering a central heating system.

Combination valve – Many areas of plumbing require a number of valves to control the water. The most common scenario uses a filter to clean the water of debris, a check valve to ensure the water only flows in one direction, and a pressure-reducing valve to keep the water pressure under control. Rather than make three separate devices all the functions are combined into a single combination valve. This is a generic term for any valve that undertakes a number of distinct tasks, although the tasks themselves can differ.

Compression fittings – A type of fitting that makes a watertight seal by compressing a ring of brass or copper between the water pipe and the fitting body.

Condensate pipework – Modern boilers cool the flue gases down so much that they condense back into water. The condensate pipework takes this water away from the boiler and into the household's waste pipework.

Condensing boiler – A modern boiler that reduces fuel bills by removing as much heat as it can from the waste flue gases. This recovered heat is then used to keep the home warm. In removing this heat some of the flue gas condenses back into water, hence the name.

Degrees Clark – A way of measuring how much calcium carbonate 'scale' is in your water, ie how hard it is: 1 degree Clark equals 14.25 parts per million of calcium carbonate. Different countries have different values for a degree clark, so check where you're living!

Double-check valve – A check valve is used to ensure that water can only run through a pipe in one direction. If the water attempting to come back up the pipe in the wrong direction is dirty, or potentially dirty, then a double-check valve is used, so that if the first check valve fails the second stops the flow.

Drain auger – A wire that's passed down a drain and then twisted in an attempt to clear a blockage.

Earthing – Most pipework in the home is made from copper, which is very good at conducting electricity. To make sure people don't electrocute themselves when they touch a pipe the pipework itself is attached to a cable that leads to the main electrical earth in the house.

Elbow joint – A plumbing fitting that allows the direction of a pipe to change by 90 .

End-feed fittings – The cheapest form of plumbing fitting, which works by applying solder to the end of a fitting and letting it run into the fitting to form a watertight seal.

Equal tee – A T-shaped fitting where all three pipes going into it are the same size.

Evacuated tube panels – A form of solar thermal panel that uses a vacuum to increase performance and provide hot water for the home even when it isn't sunny.

Expansion tank – When water heats up it expands. This tank is used to contain the expanded water. Often called the 'feed and expansion' tank as it performs two roles – feeding fresh water into the system and allowing hot water to expand safely.

Expansion vessel – When heated water can't expand its pressure increases instead. An expansion vessel prevents the pressure rising too much by giving it somewhere to expand into, thus lowering the pressure.

F&E tank – The feed and expansion tank. A small loft tank that 'feeds' fresh water into the central heating system and provides the hot water with somewhere to expand into.

Filling loop – A device used to increase the pressure of 'sealed' central heating systems, usually after one or more radiators have been bled.

Flat-panel collectors – Type of solar thermal panel that uses the sun's energy to create hot water for the home.

Full crossover – Type of fitting that allows one pipe to pass over another without the pipes touching.

G3 – Part G, section 3, of the building regulations, which describes how an unvented hot-water cylinder should be fitted and maintained. You cannot fit or service an unvented cylinder until you've completed a qualification course covering this part of the regs.

Gas Safe register – The organisation that took over from Corgi and is now responsible for gas safety in the UK. You cannot fit or service any gas appliance if you aren't Gas Safe registered.

Gate valve – Type of tap that raises and lowers a bar of metal, or 'gate'. When the gate is down the tap is closed, when it's up the tap is fully open. They're popular because when the gate is up water can flow straight through without any restrictions.

Gland nut – The outside garden tap is a classic example of a tap that uses a gland nut. This is where a waterproof material is packed down to form a watertight seal around the tap's spindle, allowing the tap to be turned on and off without water leaking out of the top.

Grey water – Waste water from your bath, shower, washbasin and kitchen sink.

Heat pump – A fridge is a type of heat pump. It uses pressures and refrigerant liquids and gases to move heat from one place to another, often magnifying the heat gain or loss at the same time. Modern heat pumps can be used to heat the home very efficiently.

Hose union backplate – A metal plate that screws to the wall and supports a hose union bib tap.

Hose union bib tap – The outside garden tap is a hose union bid tap in that it has a spout to which a hose can be attached and the tap itself is screwed into a back plate that holds it to the wall.

Immersion heater – Effectively a kettle element that fits inside a hot-water cylinder and uses electricity to heat up the water.

Inhibitor – Chemical added to a central heating system that helps to slow the rate of corrosion within the pipework, boiler and radiators.

Isolation valve – A valve that's fitted close to taps and ball valves and can be used to turn them off if they go wrong, making repair and maintenance much easier.

Jointing compound – Sticky gunk that's used to create a watertight seal on threaded joints.

Kettling – The horrendous noise made by a boiler when hot water turns into steam as a result of debris or scale building up inside it or it's associated pipework.

Lead-Lok – Generic name given to a fitting that's used to convert from lead to copper or plastic pipework. It's actually the name of a single product but is used in much the same way that all vacuum cleaners are called 'Hoovers'.

Limescale – Most water in the UK contains lots of minerals, which when the water is heated turn into a hard deposit called limescale. This sticks to the inside of kettles, pipes and boilers etc.

Lock-shield – At each end of a radiator is a valve. One is used to turn the radiator on and off whilst the other is called the lock-shield valve. This is set when the central heating system is installed and is hidden away under a little cap so that its settings can't accidentally be changed, as doing so can affect the performance of the radiators.

Machine bend – This is where you bend a pipe rather than use a fitting. A machine called a tube bender makes the bend, and this results in a much gentler bend that doesn't restrict water flow as much as a fitting and is far less likely to leak in the future.

Mains stop tap – When the mains cold water enters your home the first tap is comes across is the mains stop tap. Closing this tap turns off all the cold water in your home immediately and stops any more water entering the house.

Microbore – Many modern central heating systems use pipework only 8mm or 10mm in diameter, referred to as 'microbore'.

Monobloc tap – A tap with a single spout from which both hot and cold water emerges.

MUPVC – Type of plastic used in plumbing. It has the huge advantage of not turning brittle when exposed to sunlight and is therefore ideal for guttering and soil stacks etc. Also called PVC-C.

OFTEC – The oil-firing technical association. It operates a 'competent persons' scheme for technicians working on oil-fired appliances.

Olive – A metal ring that creates a watertight seal when crushed between a pipe and a compression fitting.

Open-flued boiler – An old-fashioned type of boiler where some or all of the air it uses to burn fuel is taken from the room in which it's situated. As a result they need ventilation and can be far more dangerous than modern 'room-sealed' boilers.

O-ring – Rubber ring used to create a watertight seal.

Pan connector – The fitting that connects the toilet to the waste pipe.

Partial crossover – Fitting used to take one pipe over another without the two touching. Differs from a full crossover in that it usually involves a pipe branching off or changing direction.

Pedestal – Most washbasins sit atop a pedestal. These support the weight of the basin and hide unsightly pipework from view.

Photovoltaic solar panels – Solar panels used to generate electricity.

Pipe clip – A clip used to support pipework, keeping it firmly in place and stopping it vibrating as water passes through it. Usually made from plastic.

Pipe cutters – A tool used to cut pipework cleanly and precisely.

Pipeslice – Another name for a pipe cutter.

Plumbers' mait – A type of putty that doesn't set. Used to create a watertight seal, and can be used instead of a washer in some circumstances.

Pop-up waste – Type of plug connected to a metal bar. The bar is usually positioned behind a tap – press the bar down and the plug opens, pull it up and the plug closes.

Potable water – Water that's fit for drinking.

Powerflush – Process that cleans out the debris from inside a central heating system.

PPE pack – A 'personal protective equipment' pack, usually consisting of protective eye glasses, a dust mask and a set of ear defenders.

Primatic cylinder – Old type of hot-water cylinder that uses nothing more than an air bubble to keep the central heating and hot water separate.

PTFE tape – Polytetrafluoroethylene. A plastic tape that can be used to create a watertight seal.

Pump pliers – An essential plumbing tool used to tighten and loosen nuts.

PV – Short for 'photovoltaic', a type of solar panel that uses the sun's energy to generate electricity.

PVC-C – See MUPVC.

Quarter-turn tap – A modern tap that only requires the head to be turned by 90° (one quarter of a circle) to go from fully closed to fully open and visa versa.

Reducer – Generic term for a fitting that connects pipework of different sizes, eg to go from 22mm pipe to 15mm pipe you'd buy a 22–15mm reducer fitting.

Regulation 16 kit – Another name for a Byelaw 30 kit.

RHI – Renewable Heat Incentive. At the time of writing this was the new grant scheme for energy saving and renewable energy technology.

Riser kit – A kit comprising numerous legs to raise a shower tray off the floor.

Rising spindle tap – The standard outdoor tap. When the tap is closed some of the spindle to which its head is attached disappears into the tap body, and when it's opened more of the spindle reappears.

Rodding kit – Kit used to clear drains – usually outdoor drains.

Room-sealed boiler – A modern boiler that takes all its air from outside, combusts air and fuel in a chamber sealed from the room housing it, and blows its fumes outdoors.

SDS drill – Special direct system. A type of drill chuck and drill that allows the user to change bits without a key.

Siphon – The gadget within a toilet cistern which, when operated by a handle or button, causes the toilet to flush.

Slip straight coupling – A pipe fitting used to connect two straight pipes when there's absolutely no give in the pipework.

Slip washer – A hard, slippery plastic washer that sits between a rubber washer and a nut and allows the nut to be tightened without ripping apart the rubber washer.

'Smart' pump – A central heating pump which automatically adjusts its speed and energy consumption according to the heating needs of the system.

Soakaway – An area underground that lets water from the gutters or a condensate pipe flow into it and slowly soak away. Usually filled with loose rubble, stones or plastic crates.

Socket – Alternative name for a straight coupling or straight fitting used to connect two pipes together without a change of direction.

Solar thermal panels – Solar panels that use sunlight to heat water, either directly or indirectly, in order to generate hot water for the home.

Speed-fit – Type of fitting used to fit copper or plastic pipework together.

Stand-off clip – A clip that secures pipework to a wall while still allowing enough room between pipe and wall for insulation to be fitted.

Stopcock – Alternative name for the mains stop tap.

Stop-end – Name given to a fitting used to terminate a pipe run either permanently or temporarily. Most often used when an outlet, such as a toilet, is moved and the old pipework is no longer needed.

Stop tap – An alternative name for the stopcock or mains stop tap.

Straight coupling – A fitting used to connect two lengths of pipe together to form one long pipe. Also known as a socket.

Street elbow – An elbow fitting where one end slides inside another fitting. This reduces the distance needed to make a 90° turn.

Superflux – Flux is used in soldering and allows the solder to flow into the fitting correctly. Superflux is a generic term for a flux that contains a weak acid that helps clean the pipework at the same time, ensuring a watertight joint.

System boiler – Boiler designed to work with in a high-pressure central heating system. It usually comes with a built-in expansion vessel, pressure relief valve, pressure gauge and pump.

Tail – Term used to describe the open end of a pipe.

Tap connector – Fitting used to connect pipework to the base of a tap.

Tectite – Name given by the manufacturer to a type of metallic push-fit fitting.

Trap – Also called a U-bend, it's a section of waste pipework that's permanently filled with water. This water creates a barrier and stops smells coming out.

Tri-tap – A tap with three handles – one for hot, one for cold and one for filtered cold water.

TRV – A thermostatic radiator valve, a radiator valve that automatically turns off the radiator once it's reached a certain temperature.

Tundish – An open, cup-shaped gap between two vertical pipes. It's a safety device that allows you to see when water is running through the pipework connected to one or more safety valves, ie when water is running through the tundish something has gone wrong.

UFCH – Underfloor central heating. A series of pipe loops under the floor which act like a radiator.

Unequal tee – A fitting that allows three pipes of differing sizes to be connected together.

Universal fitting – A waste pipework fitting that allows pipework from different manufacturers to be connected together.

Unvented cylinder – A hot-water cylinder where the water is stored at mains pressure.

Upstand – The vertical lip on the edge of a shower tray.

Vent pipe – A pipe that terminates over a tank and allows hot water to escape rather than become over-pressurised.

Vented cylinder – A hot-water cylinder where the water pressure is created by a storage tank situated above it, usually in the loft. The hot water is maintained at normal atmospheric pressure by a vent pipe that discharges over the loft tank if things go awry.

Waste pipe – Generic term for pipework that takes waste water away from a toilet, washbasin, bath, sink, shower etc.

Willis Solasyphon – Manufacturer's name for a device that can convert a standard hot-water cylinder into one suitable for use with a solar thermal panel.

WRAS – Water Regulations Advisory Scheme. The UK body that offers advice and guidance to manufacturers and consumers regarding the current water regulations.

ZNE – Zero net energy. An acronym used to describe a home with zero net energy consumption or carbon emissions. This is usually achieved via high levels of insulation, wind and solar energy sources, low-energy lighting etc.

Useful contacts

Gas Safe register – The organisation that oversees all gas safety in the UK. Gas engineers must be registered with the Gas Safe register in order to undertake any gas-related work. As such you can also use this site to find local qualified plumbers – www.gassaferegister.co.uk.

OFTEC – If you have an oil-fired stove or boiler this is a site worth visiting for a list of OFTEC-approved technicians – www.oftec.org

Environmental Agency – This agency deals with everything environmental – from septic tanks to fishing licenses, it's all here. It's also a good source of information for any grants that might be available to homeowners – www.environment-agency.gov.uk.

WRAS – The advisory body dealing with the water regulations. If you have a query on the regulations try contacting these people or your local water company – www.wras.co.uk.

DEFRA – The Department for Environment, Food and Rural Affairs. This is usually the place where news of any new grants first appears – www.defra.gov.uk.

Energy Saving Trust – Nice website with loads of information on green technologies and grants – www.energysavingtrust.org.uk.

RHI – The Renewable Heat Incentive is the current grant system available for green and energy-saving technologies – www.rhincentive.co.uk.

UPDATES – For updates and news on this manual, visit www.a1perfectplumbing.co.uk

Contacts

Your local water company is a good source of information and advice and can usually be found by just typing their name into an Internet search engine. Alternatively dig out a water bill and give them a call – although this is regarded as rather archaic these days, and they may act as if you just rode into town on a dinosaur.

If you're interested in a career in plumbing get in contact with your local college. Most run plumbing courses at NVQ 2 and 3 level and when they don't they can usually point you in the right direction.

If you're young and interested in plumbing then have a look at www.apprenticeships.org.uk for information on how to become a plumbing apprentice.

Further reading

If you're interested in a career in plumbing you should consider taking an NVQ in Plumbing. There are umpteen books that will take you though the plumbing theory as part of a college course, but they're universally heavy going and filled with as much wit and humour as a morgue. That said, they go through things in far more detail than we can here and, if nothing else, will give you an insight into what you're letting yourself in for. All are available via well-known online bookshops.

Index

abrasive strips 35, 40, 62
accumulators 19
acrylic baths 146-8
actuators 191
adjustable spanners 32
airlocks 107-8
air-source heat pumps 213-14
alternative fuels 203, 209, 215-16
asbestos 32

backflow 13, 116
back-to-the-wall toilets 139-40
balancing central-heating systems 186, 188
ball valves
  failing 38, 80
  frozen 42-3
  leaking 85
  replacing 98-102
  in storage tanks 21, 23-4, 50, 101-2, 109, 178
  in toilets 92, 94, 97-100, 138, 142
basin fixing kits 162-3
basin spanners 33, 135
bath legs 147, 148, 152-3, 156
bath panels 146, 150
bath taps 135-6, 147, 149, 153-4, 156
bathroom suites 133
baths 135-6, 146-57
bend supports 70
bending pipework 70-2
bidets 169-70
biomass boilers 215-16
bleed valves 28, 173, 177, 180-1, 184
bleeding radiators 28, 173
blending valves 205
blocked drains 89-90
blocked sinks 85-7
blocked toilets 88
blowtorches 60, 61-2
boilers 25-6, 43, 188-9, 199-201, 215-16
bottle traps 86, 161, 163
box spanners 33-4, 119
Bramah, Joseph 10
branches (in pipework) 47-8
buffer tanks 205
building regulations 10-11

bungs 50
butterfly nuts 95, 97, 138, 143
Byelaw 30 kits 48-51, 109

captive-plug waste mechanisms 160
carbon monoxide poisoning/detectors 10, 200
cast iron baths 146, 147
cavity wall insulation 203, 206
central heating systems
common problems 186-9
  draining 27-9
  leaks in 37, 39
  maintaining 173-8
  powerflushing 174-7
  refilling 29
  types of 23-6
  underfloor (UFCH) 203-9
central-heating pumps 186-7, 189-90
chrome pipework 60, 68, 165
cisterns (toilet) 91-4, 96-7, 137-43, 145-6
clamp repair kits 39
'clicker' waste mechanisms 160
close-coupled toilets 137, 139
'coffin' tanks 109
cold-water storage tanks
  function of 18-19
  and hot-water stop tap 78-9
  identifying 21-2
  insulating 48-51
  overflowing 101-2
  replacing 108-11
combi boilers 20, 187-8, 201, 211
combination valves 20
common sense 32
compression fittings 56-9, 67, 83-4
compression repair kits 42
condensation 83
condensing boilers 20, 201
contamination 13, 116
COP (coefficient of performance) values 213, 215
cordless power tools 33
core drill sets 125, 159
corners (in pipework) 46-7

Crapper, Thomas 9-10
cutting pipework 34-5, 53, 66

'degrees Clark' values 131
digital showers 169
dishwashers 124-8
diverter valves 187
DIY emergencies 38
double glazing 203, 206
double-check valves 13, 116, 170
doughnut seals 94-7, 142
drain augers 88
drain cleaner 86, 89
drain cocks 27-8, 76, 119, 194, 195, 197
drain covers 89
drains (blocked) 89-90
drills 33
dripping taps 102-7
Dudley Turbo siphons 93, 95
dust masks 32

ear defenders 31
earthing pipework 40, 42, 53
elbows (in pipework) 68
electric showers 165, 169
electric underfloor heating mats 204
electric water conditioners 130
emergencies 37-43, 75
end-feed fittings 62-3
energy neutral homes 203
equal tees 69
erroneous measurement 13
evacuated tube collectors 210
expansion vessels 25, 188-9
explosions 10-11, 192, 199

fat 85
feed and expansion (F&E) tanks 23-4, 25, 49, 177, 178
filling loops 25-7, 29, 131, 188
fire extinguishers 61
fitting baths 147-57
fitting radiators 181-5
fitting toilets 139-46
fitting washbasins 158-63
flat-panel collectors 209-10
flexible tap connectors 120-1, 145, 154-6, 161

flow detectors 187
flow restrictors 142
flush pipes 137-8
flushing mechanism (toilet) 91-7, 137, 141, 145-6
flux 61, 62-3
foam insulation 45-8
foil-backed insulation 45
fortic cylinders 18
fossil fuels 203, 215
free-standing baths 147-8
free-standing washbasins 158
frozen condensate pipework 43
frozen pipes 37, 41, 42-3, 45
full-bore lever valves 78-9, 80, 111, 167, 168

gardens 113-17
Gas Safe register 10
gate valves 18-19, 24, 78-9
gland nuts 75, 77, 188
global warming 203
gravity mixer showers 166-7
gravity-fed central heating systems 23-4, 178
grey water harvesting 219
ground-source heat pumps 214-15

hacksaws 35
hammer drills 33
Harrington, John 10
Hawksley, Sir Thomas 10
heat mats 61
heat pumps 212-15
heated towel rails 185
Hep2O push-fit system 55
hessian roll 45
hiding pipework 164
high-pressure central heating systems 25-7, 188
high-pressure hot-water systems 19-20
high-pressure mixer showers 167
history of plumbing 8-10, 53
holidays 15, 22-3, 37
hose connectors 57
hose union backplates 114
hosepipes 13

hot water cylinders
  fitting 192-7
  primatic 18, 174
  solar powered 211-12
  unvented 10-11, 19-20, 78, 192
  vented 11, 193
hot-water stop taps 78-9, 108
hot-water systems 17-20

ice-makers 128
immersion heaters 193-4, 197-9
In4sure push-fit system 55
inhibitor 177-8
injuries 7, 31-2
inset washbasins 158
insulating cold-water storage tanks 48-51
insulating pipes 37, 45-8
insulation 203, 206
integral solder ring fittings 61-2
ion replacement water softeners 130-1
iron and steel pipework 63-4
irrigation systems 116
isolation valves
  on bath taps 147, 149, 156
  on central-heating pumps 189-90
  on cold-water storage tanks 21, 24, 111
  on feed and expansion tanks 177
  fitting 80-81
  function of 79-80
  on kitchen taps 119
  on outside taps 115
  on showers 165, 167
  on test rigs 57
  on toilets 137, 145
  on washbasins 159, 163
  on washing machines 126

jigsaws 122
jointing compound 35, 58, 64, 77, 196
jubilee clips 125, 127

kitchen sinks 121-4
knee pads 31

laminated flooring 205
lead pipework 53, 64-5, 129
Lead-Lok fittings 65
leak repair tape 39
leaks
  causes of 37-8
  in compression joints 83-4
  detecting 83
  repairing 39-42, 83-5
lever valves 78-9, 80, 111, 167, 168
limescale 64, 80, 104, 130, 186
lock-shield valves 177, 179
low-level UFCH 206-7
low-pressure hot-water systems 18-19

machine bends (in pipework) 48
magnetic cleaners 178
magnetic water conditioners 130
main drain covers 89-90
mains stop tap 15, 75-7
manhole covers 89-90
manifolds 205, 206-9
masking tape 121-2, 144
microbore copper tube 70-1
mixer showers 108, 164, 166-9
mixer taps 103
monobloc taps 102-3, 105, 119-21, 160-1
motorised valves 175-7, 187, 191-2

nails (in pipework) 38
negative head pumps 168
noisy central-heating systems 186

oil boilers 10
olives 56, 58-9, 60
open gully drains 89
open-flued boilers 200
outdoor pipework 113
outdoor stopcocks 15-16, 75
outside taps 113-16
overflow pipes
  in baths 154-5
  in cold-water storage tanks 21, 24, 49, 101, 109, 111

failing 38
  in kitchen sinks 123
  in toilets 94, 96, 138
  in water butts 117
overflowing toilets 97-100

paint removal (from pipes) 81
pan connectors 88, 138, 143
partial crossovers 69
pedestal washbasins 159-63
phosphate water conditioners 130
photovoltaic (PV) solar panels 215
pipe benders 70-2
pipe clips 70, 73
pipe cutters 34, 53, 66
pipework
  bending 70-2
  branches in 47-8
  burst 38-9, 83
  chrome 60, 68, 165
  compression fittings 56-9
  corners in 46-7
  cutting 34-5, 53, 66
  earthing 40, 42, 53
  elbows in 68
  fittings 68-9
  frozen 37, 41, 42-3, 45
  hiding 164
  insulating 37, 45-8
  iron and steel 63-4
  lead 53, 64-5, 129
  machine bends 48
  outdoor 113
  overflow see overflow pipes
  paint removal from 81
  push-fit 53-5
  repairing 39-42
  sizing 72-3
  soldered 60-3
  solvent weld 66-7
  supporting 73
  underground 70
  waste see waste pipework
planning 32
plastic push-fit pipework 53-5
platform construction (water tanks) 110
plumbers' hemp 58
plumbers' mait (putty) 95, 123, 160

plungers 85, 88, 138
pop-up waste mechanism 155, 160
power showers 169
powerflushing central heating systems 174-7, 187
PPE (personal protective equipment) packs 32
pressure gauges 25-6
primatic hot water cylinders 18, 174
printed circuit boards (PCBs) 201
profile tools 133, 135, 156
PTFE tape 35, 58, 64, 75, 84-5, 182, 196
p-traps 86-7, 155-6, 161, 163
pump pliers 32, 81, 194-5
pumped electric showers 169
push-button waste mechanisms 136
push-fit pipework 53-5
push-fit repair kits 39-42
push-fit stop-ends 34-5
push-fit tank connectors 110
push-fit waste pipework 65-6
PVC-C plastic 67

quarter-turn taps 102, 104-5

radiator tails 180-5
radiators
  bleeding 28, 173
  common problems 186-8
  and draining central heating systems 28-9
  fitting 181-5
  powerflushing 174-7, 187
  removing 179-81
radon gas 203
rainwater harvesting 216-19
reducers 69
Regulation 16 kits 48-51, 109
removing baths 146-7
removing radiators 179-81
removing toilets 137-9
removing washbasins 157
RHI (Renewable Heat Incentive) grants 209, 219
riser kits 170-1
rising spindle taps 102
rodding kits 90

rodents 37, 45, 53, 55, 165
rogue tradesmen 10
room-sealed boilers 200-1

safety equipment 31-2
screed floors 205-6, 207, 209
screwdrivers 33
screws (in pipework) 38
sealant 122-3, 133, 135, 145, 156
self-levelling compound 171
self-sealing traps 156, 162
semi-recessed washbasins 158
sewerage systems 8-9
shower baths 148
shower pumps 167-9
shower rails 164, 165
shower trays 134, 170-1
showers 108, 164-9
silicon grease 66, 107
sink clips 123
sinks (blocked) 85-7
siphons (toilet) 91-7, 100, 141-2, 145-6
sizing pipework 72-3
slip straight couplings 68
slip-sockets 42
slip-washers 160
'smart' pumps 189
soil stacks 149-50, 156, 159, 171, 219
solar cylinders 211-12
solar thermal energy 209-12
solder 61, 62-3
soldered pipework 60-3
soldering irons 60, 62
solid wood flooring 205
solvent weld pipe 66-7
speed-fit system 53-4
s-plan systems 191
stand-pipes 125, 127-8
steel baths 146, 147
stilson wrenches 63
stop-ends 69, 127
stored systems 17
straight couplings 68
s-traps 86-7, 161

street elbows (in pipework) 68
strength 7, 34
stud detectors 150-1
superflux 62
super-seal pipe inserts 53-4
supporting pipework 73
swan-neck taps 106-7
system boilers 201

tap connectors 57, 69, 120-1, 145, 154-6, 161, 162-3
tap reseating tool 105
tap restorer kits 106
tape measures 33
taps
    bath 135-6, 147, 149, 153-4, 156
    dripping 102-7
    hot-water stop 78-9, 108
    identifying 15-16
    mains stop 15, 75-7
    monobloc 102-3, 105, 119-21, 160-1
    mixer 103
    outside 113-16
    quarter-turn 102, 104-5
    rising spindle 102
    swan-neck 106-7
    'tri' 129
    washbasin 135-6, 157, 160-1
tectite metal push-fit pipework 55
tee fittings 69, 115
tennis elbow 7, 32
test rigs 56-8
thermal store systems 20, 192, 205
thermostatic radiator valves (TRVs) 179-80, 184-5, 188
thermostats 198, 199, 208
tile drills 145
tiling 133, 134, 171
toilet seats 146
toilets
    blocked 88

fitting 139-46
flushing mechanism 91-7, 137, 141, 145-6
invention of 9-10
overflowing 97-100
and rainwater harvesting 216-19
removing 88-9, 137-9
tools 32-5
traps 85-7, 147, 155-6, 161-3
tray formers 171
'tri' taps 129
tundish 11
twin impeller pumps 168

underfloor central heating (UFCH) 203-9
underground pipework 70
undue consumption 12
unvented hot-water cylinders 10-11, 19-20, 78, 192
upstands 170
utility knives 33
UV degradation 67

valves
    ball see ball valves
    bleed 28, 173, 177, 180-1, 184
    blending 205
    combination 20
    diverter 187
    double-check 13, 116, 170
    gate 18-19, 24, 78-9
    isolation see isolation valves
    lever 78-9, 80, 111, 167, 168
    lock-shield 177, 179
    motorized 175-7, 187, 191-2
    thermostatic radiator (TRVs) 179-80, 184-5, 188
    wheel head 179
vent pipes 11, 21, 24, 50-1, 111, 196
vented hot-water cylinders 11, 193

ventilation 200
vinyl flooring 171

washbasins 135-6, 157-63
washers 35, 76, 84-5, 102-5, 136
washing machines 124-8
washing-machine connectors 124, 126-8
waste pipework
    baths 149-50, 154-6
    kitchen sinks 123-4
    showers 170-1
    toilets 138-40, 143
    types of 65-8
    unblocking 85-7
washbasins 159, 161-3
washing machines 125, 127
wasting water 12
water butts 117, 218
water conditioners 130
water filters 129
water meters 13, 16, 117, 216
water regulations 10, 12-13
Water Regulations Advisory Scheme (WRAS) 12
water softeners 129-31
water vacuum cleaners 35, 49, 86, 107, 138
waterborne diseases 9
WD40 oil 76
'wet' central-heating systems 23
wet rooms 171
wet underfloor heating systems 204-9
wheel head valves 179
Willis Solasyphon system 211
wood-burning stoves 215-16
work trousers 31
worktops 121-2

Y-plan systems 191

zero net energy (ZNE) buildings 203
zinc water conditioners 130